Scholastic Culture in the Hellenistic and Roman Eras

Transmissions

Studies on conditions, processes and dynamics of textual transmission

Edited by
Rosa Maria Piccione

Volume 2

Scholastic Culture in the Hellenistic and Roman Eras

Greek, Latin, and Jewish

Edited by Sean A. Adams

DE GRUYTER

ISBN 978-3-11-065787-6
e-ISBN (PDF) 978-3-11-066098-2
e-ISBN (EPUB) 978-3-11-065799-9
ISSN 2625-4018

Library of Congress Control Number: 2019947085

Bibliographic information published by the Deutsche Nationalbibliothek
The Deutsche Nationalbibliothek lists this publication in the Deutsche Nationalbibliografie; detailed bibliographic data are available on the Internet at http://dnb.dnb.de.

Cover image: Salone Sansovino © Biblioteca Nazionale Marciana
Printing & binding: CPI books GmbH, Leck

www.degruyter.com

Acknowledgements

This project would not have been possible without the generosity of the British Academy. The volume before you is a collection of essays presented at 'Ancient Scholarship: Scholastic Culture in the Hellenistic and Roman Eras', held at University of Glasgow in April 2017. This was the follow-up conference to a colloquium in October 2015 on 'The Future of Academic Publishing', also held at the University of Glasgow. Both events and the subsequent dissemination of findings are in fulfilment of the British Academy's Rising Star for Engagement Award that I received. I am very grateful to the Academy for their investment in early career scholars and their willingness to fund topics that might not otherwise receive financial support.

Thanks are due to my institution, the University of Glasgow, and my colleagues in both the Classics and Theology and Religious Studies subject areas. My colleagues make Glasgow University a supportive place to work and I appreciate that departmental categorisations are not a barrier to collaborative work. Thank you to all who attended and presented.

I would also like to thank the different reviewers of this volume and their comments, which helped make the volume better. Zachary Vickery, my doctoral student, is also to be thanked for his work on the indices.

Finally, a majority of the organisational work for both conferences was undertaken by Zanne Domoney-Lyttle. Zanne worked tirelessly to ensure that the events progressed smoothly and that the presenters were well plied with coffee, scones, and Scottish food. Both conferences are better because of her involvement and I am deeply grateful for her efficiency, positive attitude, and support.

Sean A. Adams
Glasgow, UK
March 2019

https://doi.org/10.1515/9783110660982-001

Contents

Sean A. Adams

Introduction: Themes in Ancient Scholarship

This volume is a collection of papers presented at 'Ancient Scholarship: Scholastic Culture in the Hellenistic and Roman Eras', a conference funded by the British Academy and held at the University of Glasgow, UK on 27–28 April 2017. The purpose of this conference was to investigate scholastic culture in the Hellenistic and Roman eras, with a particular focus on ancient book and material culture as well as scholarship beyond Greek authors and the Greek language. Although ancient scholarship is a burgeoning field of study, a majority of investigations have focused primarily (and sometimes exclusively) on Greek scholarship and do not readily engage with the literature and languages of other cultures.[1] Such studies are valuable contributions, but they do not represent the breadth of scholastic activity undertaken in antiquity. Accordingly, one of the major contributions of this work is the inclusion of multiple perspectives and its contributors engage not only with elements of Greek scholastic culture, but also bring Greek ideas into conversation with developing Latin scholarship (see chapters by Dickey, Nicholls, Marshall) and the perspective of a minority culture (i.e., Jewish authors) (see chapters by Hezser, Adams). This multicultural perspective is an important next step in the discussion of ancient scholarship and this volume provides a starting point for future inquiries.

In particular, the essays in this volume make important contributions and provide fresh insight into a few, specific areas. First, one of the fundamental aspects of this collection is the importance of book culture and the materiality of scholarship. Many works on ancient scholarship are limited to discussions of specific authors, texts, and ideas, but do not take into account the means by which scholarship is accomplished in ancient times. This singular focus is a clear oversight of the highly complex and interconnected nature of scholastic research. As a result, the importance of the intersection of text and scholar is highlighted in a number of chapters and is one of the theoretical underpinnings of this volume.

This framework is evident in the chapter by Stephanie Roussou, 'New Readings in the Text of Herodian'. Although Roussou provides fresh proposals for Herodian's Περὶ καθολικῆς προσῳδίας (*On Prosody in General*), her discussion is not limited only to the text as reconstructed. Rather, Roussou also evaluates newly-discovered manuscripts and discusses what they mean for our understanding of Herodian, his work and its reception, and the scholastic environment in which he wrote and in which

1 For traditional scholarship, see, e.g., Sandys (1903–1908); von Wilamowitz-Moellendorff (1921); Pfeiffer (1968). For more recent examples, see Dickey (2007); Montanari/Perrone (2009); Matthaios, *et al.* (2011); Montanari, *et al.* (2015). For engagement with other cultures, see Honigman (2003); Niehoff (2011).

Sean A. Adams, University of Glasgow

https://doi.org/10.1515/9783110660982-002

his work was used. Here Roussou takes seriously the material nature of scholastic production and provides new and corrective insights to previous scholarship.

Fundamental to discussions of scholarship is the question of the scholar's access to texts. When thinking about scholarship in the Graeco-Roman world it is tempting to assume that scholars have ready access to the texts and materials necessary for their study. However, this assumption is rarely accurate, not only amongst those in communities and cultures that are not known for being intellectual hubs, but also amongst Greek and Latin elites. As can be seen from the laments of elite authors in the Roman era, even prominent scholars and writers have difficultly accessing texts. Moreover, the texts to which they do have access are not always in great condition or well edited (e. g., Galen, *Peri Alupias* 13, 19; Plutarch, *Dem.* 1.1–2).[2] Christian and Jewish authors, although primarily focusing their intellectual energy on their authoritative religious texts, do not always have direct access to them.[3] It might be that the texts that they do have are communally owned and so are not regularly or openly available (e. g., in a prayer house/synagogue/religious building). In addition, the texts that they do have may not be complete (i. e., a community might not have been able to acquire/afford a copy of all scriptural texts or every scroll of a large work). Similarly, when writers and scholars are travelling throughout the Greek, Roman, and Syrian worlds, something that is increasingly being recognised as a common phenomenon,[4] the bulky and cumbersome nature of texts precludes their accompaniment. In these locations and situations, memorisation of texts plays a much more important role both in educational programmes as well as in scholarly discourse.[5]

At the same time, there is a strong, and indeed necessary, connection between texts and oral tradition. Not all scholarship is limited to texts, as can clearly be shown in some of the philosophical and rabbinic schools.[6] In these communities, texts as well as received teachings form the basis of scholarly discussions, but they are also subsequently commented upon by later teachers, whose teachings, if warranted, become codified and part of the received tradition for subsequent generations. The intersection of oral and written teaching and the deep interplay between them through the centuries is understudied, primarily because we only have access to oral material if it was written down. In this instance, Classics scholars have much to gain from scholars of rabbinics and of later Judaism, whose key texts (i. e., Qumran

2 For discussions of personal libraries and access to texts, see, for example, Houston (2014).

3 Cf. Albl (1999); Albl (2016).

4 E. g., Tacoma (2016); Isayev (2017).

5 For the role of memorisation in Greek and Latin education, see, for example, Asmis (2001) 216–22. For the way that rabbis interpreted the lack of access to texts as an advantage, see y. Peah 2:6, 17a and the discussion by Hezser in this volume.

6 Oral tradition is evident in Greek philosophical communities (e. g., the disciples of Pythagoras) as well as rabbinic documents (e. g., Mishnah and Talmud). For a discussion of oral education and scholarship in the Qumran community, see Lemaire (2006); Brooke (2017) 21–26. Cf. Cooper (2007).

scrolls, Mishnah, Tosefta, targumim) are codifications of collected wisdom of previous sages and represent one aspect of the process of textualisation.

When scholars do have access to texts, their location needs to be considered. In their contributions to this volume, Matthew Nicholls, Myrto Hatzimichali, and Gaëlle Coqueugniot address how literary *fora*, such as bookshops, libraries, and archives, influence ancient research practices, highlighting the issues and challenges of accessing and disseminating texts in the Graeco-Roman world.

The chapter by Myrto Hatzimichali, 'Circulation of Lexica in the Hellenistic and Early Imperial Period', teases out the movement and distribution of lexica and how these scholastic works (or parts of them) left the confines of libraries and became available to a growing number of individuals. Libraries, such as those at Alexandria, Pergamum, and Rome, still functioned as the hub of technical scholarly endeavours, and as Hatzimichali rightly notes, the lexica of Zopyrion, Dorotheus, and Pamphilus, the latter of which comprised ninety-five books,[7] were unlikely to be found anywhere else. Similarly, other technical works were likely limited to major libraries (or maybe even just one library), but this did not stop portions, extracts, or epitomes of these works to leave the confines of the library and enjoy a distinct history of their own.[8] These additional 'lives' of scholarly texts are insightful when considering scholastic processes and subsequent citations or references to major works.

Matthew Nicholls, in his chapter on '"Bookish Places" in Imperial Rome: Bookshops and the Urban Landscape of Learning', highlights the blending of private and public space with regard to the book trade and intellectual discussions on available texts, moving the conversation of intellectual pursuits out of the standard institutions of learning and into the public sphere. Nicholls does not neglect the importance of libraries as institutes for study and discussion, but, though a detailed understanding of Rome's topography, he argues that neighbouring spaces with literary connections (e.g., book sellers along the *Vicus Sandaliarius*) could also be locations of dialogue and debate. The interaction of individuals within the texts discussed by Nicholls paints a picture of an urban landscape populated by a range of vested parties for whom the book trade and corresponding scholarship created a complex setting for social networking and public displays of erudition.

This larger discussion on access to books and composition lists is outworked in depth in the essay by Richard Marshall, 'Suetonius the Bibliographer'. Here Marshall investigates Suetonius' research and writing practices, specifically, his inclusion (and exclusion) of book titles in his *Vitae Caesarum* and *De grammaticis et rhetoribus*. These works, although informed by Suetonius' role as *a bibliothecis*, are not read through that lens. Rather, Marshall contrasts Suetonius' biographies, especially his restraint in cataloguing of the works of his subjects, to those of his predecessors, es-

---

7 The number is debated, ranging from 94/95 (*Suda* O 835; Π 142) to 405 (Δ 1140). Most scholars adopt 95.

8 E.g., Epitome of Pamphilus' *Glossai* by Vestinus, surnamed Julius (*Suda* O 835).

pecially Callimachus and his *Pinakes*. Taking into account both the geographical and societal situatedness of Suetonius (i.e., Rome), Marshall argues that Suetonius' work is a product of his location and that its author successfully exploited the malleable genre of biography.

Serena Ammirati investigates the nature of the scroll/codex and how material culture influenced text production. Ammirati, in 'For a Typology of the Ancient Latin Legal Book', outlines the development of the Latin legal book, arguing that a definitive typology did not emerge until the 5th century CE. Nevertheless, there are certain distinctive features (e.g., rubrics, *capita*, red ink, etc.) that show that the concept of the legal book was in the process of development.[9]

A second contribution is the importance of seeing a clear link between scholarship and education. Although there have been important works on education published recently,[10] insufficient attention has been given to this association. At its most fundamental level, education provides knowledge of culture and the ability to access and interpret authoritative texts, such as Homer, Virgil, the New Testament, Hebrew Scriptures (i.e., Torah), etc. Indeed, one of the most important aspects of scholarship, as discussed in this volume, is its textuality. Knowledge of material, which regularly comes from texts, is a central component of what constitutes scholarly practice and the ability to handle and authoritatively interpret a text is what leads to a recognition of expertise. The concept of being an expert, as well as mastering a particular body of literature, linguistic phenomenon, or cultural perspective, is a prominent theme in many of the chapters.

Eleanor Dickey, in her chapter 'What Does a Linguistic Expert Know? The Conflict Between Analogy and Atticism?', argues that expertise is a prized possession within the scholarly world and the ability to handle rare words and constructions with ease gave the one who had attained this ability prestige in elite circles. Similarly, the ability to 'translate' literary texts and to communicate between linguistic codes is also recognised as falling within the purview of the expert and not something that was within the grasp of one lacking substantial education. The acquisition of expertise through extensive Greek or Hebrew *paideia* created bonds amongst its holders, and, as Catherine Hezser argues, this affiliations results in the creation of insider and outsider categories. The topics of expertise and the acquisition of skills that would be recognised by scholastic peers are also developed in the works of

---

9 A similar argument was advanced in the conference by Zachery Cole, who argued that the earliest Christian communities, although interested in texts, did not have a regulated means by which to ensure the fidelity and uniformity of texts (i.e., the absence of scriptoria). This lack of centralised control resulted in a diversity of manuscript forms and textual variants, but the widespread use of distinctive scribal habits (e.g., *nomina sacra*, codex format, readers' aids, etc.) militates against the view that there was no continuity in copying practices.

10 Marrou (1956); Bonner (1977); Morgan (1998); Cribiore (2001); Hezser (2001); Too (2001); Bloomer (2011).

Adams (regarding translation) and Hatzimichali (regarding lexicography and linguistic specialisation).

A third contribution is the recognition that scholastic influence is not unidirectional. Although Greek practices influenced the scholarly habits and literary approaches of newly-conquered peoples,[11] they in turn were also influenced by those whom they conquered (although possibly to a lesser extent). As a result, Greek scholastic culture needs to be understood within its wider historical and geographic contexts with an eye towards identifying distinct practices and shared perspectives. For instance, Gaëlle Coqueugniot argues in her chapter that the privileging of archival material in Greek historiography and its citation as a definitive source appears to develop as Greek authors encountered scholars from the East (e. g., Babylonians, Jews, Egyptians, etc.). Authors from these cultures, especially evident in the works of Berossus, Manetho, and Josephus, place a premium on archival records as the true source of distant times and cultural heritage. Indeed, the lack of Greek record-keeping was seen to be a detriment to their ability to write history with confidence and accuracy.[12] Accordingly, Coqueugniot shows that there is a strong rise in archival use, indicating the adoption of foreign practices by Greek historians. Parallel scholarly development is also apparent in the widespread practice of commentary writing in which a lemma from a source text is commented on by an attentive reader(s). Such texts are not only found in Greek, but are also a prominent genre form in the Scrolls found at Qumran.[13]

In his chapter, 'Translating Texts: Contrasting Roman and Jewish Depictions of Literary Translations', Sean Adams undertakes a comparative work between Roman and Jewish interactions with Greek literary culture. For each group the importance of the source text, the direction of translation, and the intended purpose of the new work are key determiners of the nature of translation undertaken by the translator. Looking specifically at how literary translation is depicted in Roman and Jewish texts, Adams argues that many of the differences between Roman and Jewish authors result from the differences in power dynamics with Greek culture as well as the

11 In some instances Greek influence appears to be dominant, such as the development of ζητήματα or *quaestiones* literature (e. g., Aristotle's *Poet.* 25 1460b6 – 1461b25; Ἀπορήματα Ὁμηρικά, Diogenes Laertius, 5.26). Gudeman (1927). See also more recent investigations, though all have strong ties to Gudeman: Gartner (2002); Dreyer (1975). A similar type of format (i. e., ἐρωτήσεις καὶ ἀποκρίσεις) is found in select educational texts (e. g., Stephani version of the *Hermenumata*). See Dickey (2012 – 2015), 1.26 – 27. This form of literature was adopted by the Jewish author Philo of Alexandria in his *Questions and Answers on Genesis and Exodus* and there is evidence of its inclusion in Demetrius the Chronographer's *History* (fr. 1.14; fr. 2.1 – 3; fr. 5.1).

12 Cf. Josephus, *C. Ap.* 1.19 – 22; Tertullian, *Apol.* 19.5 – 6.

13 E. g., Greek: Derveni Papyrus; Hipparchus of Nicaea's commentary on Aratus' *Phaenomena*; *Anonymous Theaetetus Commentary* (*P.Berol. inv.* 9782); Philo of Alexandria's *Allegorical Commentaries*; Hebrew *pesharim* (1Q14 – 1Q16; 3Q4; 4Q161 – 4Q173; 4Q252 – 4Q254; 5Q210). For a recent, comparative study of Hebrew and Greek commentary practices, see Hartog (2017). For a discussion of commentary scholia from a range of perspectives, including the ancient Near East, see Aufrère, *et al.* (2014).

translator's relationship to the source text. By contrasting Roman and Jewish practices, Adams highlights distinctive elements and demonstrates how individuals from diverse but interacting cultures can differ in the ways in which they engage in the scholarly practice of literary translation.

In her chapter, 'Rabbis as Intellectuals in the Context of Graeco-Roman and Byzantine Christian Scholasticism', Catherine Hezser evaluates the ways that ancient Greeks and Romans marginalised Jewish Rabbis from intellectual discussions, seeing them as outside of the guild due to their communal focus and lack of Greek *paideia*; a picture that parallels the modern practice of isolating academic disciplines. Looking at the ways that Rabbis are depicted and presented in the surviving literature, Hezser argues that Rabbis saw themselves as scholars with circles of students who resembled Greek, Roman, and Christian scholars. These depictions, therefore, act as a counter-narrative to the dominant discourse and represent the claims of a minority culture attempting to navigate intellectual waters. At the same time, Hezser identifies parallel scholarly practices, particularly with regarding to encyclopaedism, reinforcing the interconnected nature of scholarship in antiquity.

In conclusion, a definition of scholarship in antiquity, including who does and does not engage in this pursuit, is of fundamental importance for how modern and ancient scholars create interpretive categories. Both ancients and moderns shape, and in many ways, presuppose the nature of 'scholarship' that is acceptable, recognised, or valid through the use of categories: a) by defining scholastic activity in a particular way (e.g., solitary versus communal; formal versus informal; intellectual versus practical); b) by use of a particular language (e.g., Greek or Latin versus Hebrew, Aramaic, Egyptian, etc.); c) by its nature (e.g., written versus oral); d) by geographic location (e.g., Rome or Alexandria versus Jerusalem, Babylon, or other 'minor' city; Roman Empire versus Syria or Babylon); and e) by outputs (e.g., commentary versus dispute genres; encyclopaedia versus law codes; books or treatises versus embodied living). Such categorical exclusions or prioritisations are outdated and have failed to provide a robust and rounded picture of scholarly practices.

The chapters in this volume have made important contributions to weakening the above dichotomies. The materiality of scholastic practice (e.g., libraries, archives, scriptoria, manuscripts, etc.), the connection between education and scholarship, and the similarities and differences amongst scholars of different communities (i.e., Greek, Roman, Jewish, Egyptian) all open new avenues of exploration regarding ancient scholarship. The purpose of this volume is to begin the conversation, one that will (hopefully) be engaged by others.

Gaëlle Coqueugniot

Scholastic Research in the Archive? Hellenistic Historians and Ancient Archival Records

The use of archival records to document research is considered common practice for many contemporary scholars in fields such as history, sociology, ethnography or geography. The archival mentality of scholarly research has, however, long been considered as a gradually developed custom, only used by the odd, ancient scholar without systematisation until modern times. Far from using archival records, ancient Greek scholars seemingly focused most of their research on relating and collating literary accounts from their predecessors and oral testimonies. Only occasionally did they refer directly to inscriptions, and even more rarely to unpublished records preserved in public or private repositories. The Hellenistic period often appears, in this context, as a shifting point between orality and literacy: documents became more frequent and better recognised, and written testimonies thus appear more readily accessible in the Greek world.

This chapter will question the scarcity of reference to archival records by ancient Greek scholars, focusing primarily on the historians of the Hellenistic period. It will begin with an overview on archives in the Hellenistic world, their characteristics and their contents. The question of the consultation of these archives by Hellenistic historians will then be addressed through several examples from the 4th to 2nd centuries BCE. The first case focuses on Polybius' famous condemnation of the historical method of his predecessor Timaeus of Tauromenion, who was known in antiquity for his careful and extensive research in Athenian archives and libraries. We will then look at the interaction between scholastic research and the archives in late Classical and early Hellenistic Athens, precisely the city where Timaeus is said to have written most of his works. A third case will focus on newly integrated regions of the Hellenistic world, with ancient civilisations where the keeping of archives was already well attested for centuries. Did this different context lead to a different use of archival records by scholars writing the history of these non-Greek civilisations? In the last part we will come back to Polybius and examine more closely his own relation to documents, specifically in a context of global history instead of a more focused chronicling on local or regional history. These four case studies will then be brought together to draw a tentative conclusion on the attitude of Hellenistic historians towards archival records.[1]

1 I am very grateful to Sean Adams for his invitation to present this research during a conference at Glasgow in April 2017, for his attentive reading of the drafts of this chapter and for his useful

Gaëlle Coqueugniot, CNRS, Maison Archéologie Ethnologie, Paris Nanterre

https://doi.org/10.1515/9783110660982-003

Archives in the Hellenistic World

Today, the term 'archives' is often taken in a somewhat restrictive meaning, as 'a collection of historical documents or records providing information about a place, institution, or group of people'[2] or the place where they are kept. The American Society of Archivists provides us with a technical definition showcasing the historical value given to archives, which constitute only the small part of the records produced by an individual or an institution that are considered of permanent value and thus worth keeping.[3] On the contrary, the French national school of archives adopts a broader definition, including all records produced or received by a person or an institution, whatever their date, their shape, or their nature.[4] Both definitions, however, emphasise two central properties of archives: their value for the individual or the institution—this can be a historical value, or more practical, legal, social or economic values—and their voluntary preservation for a varying length of time in a more or less formally organised repository, also called an *archive* or *record office*. By definition, archival records are also to be distinguished from published documents such as inscriptions and texts widely reproduced for the book market.

In the ancient Greek world, archival repositories keeping public records are attested from the Classical period at the latest. The largest and best known of these public repositories is the *Metroon* in Athens. It is mentioned by inscriptions, Athenian orators and local historians as the official repository for the archives of the *Boule* and the *Ekklesia* since its foundation in the late 5th century BCE.[5] In the Hellenistic period, there is growing epigraphic and literary evidence indicating that public record offices were a common feature in most, if not all Greek cities.[6] In addition, the preservation of archives in several private contexts is also attested, for example

---

comments. I also wish to thank the participants and audience of the conference for enriching discussions, and Julien Monerie for his insights on the cultural context of Hellenistic Babylon and Berossus' work. All opinions and mistakes remain my own.

2 *Oxford dictionary*, accessed 19 July, 2017. https://en.oxforddictionaries.com/definition/archive.

3 'Materials created or received by a person, family, or organization, public or private, in the conduct of their affairs and preserved because of the enduring value contained in the information they contain or as evidence of the functions and responsibilities of their creator [...] permanent records'. Society of American Archivists' website, accessed 19 July, 2017. https://www2.archivists.org/glossary/terms/a/archives.

4 'Ensemble de documents, quels que soient leur date, leur forme et leur support matériel, produits ou reçus par toute personne physique ou morale [...] dans l'exercice d'une activité'. École nationale des Chartes, *Dictionnaire des archives. De l'archivage aux systèmes d'information* (Paris: AFNOR, 1991), 32.

5 Sickinger (1999) 105 – 22. Several of these epigraphic and literary *testimonia* on the *Metroon* and its role as a public archive are collected in Wycherley (1957) 150 – 60.

6 Coqueugniot (2013) 29 – 37.

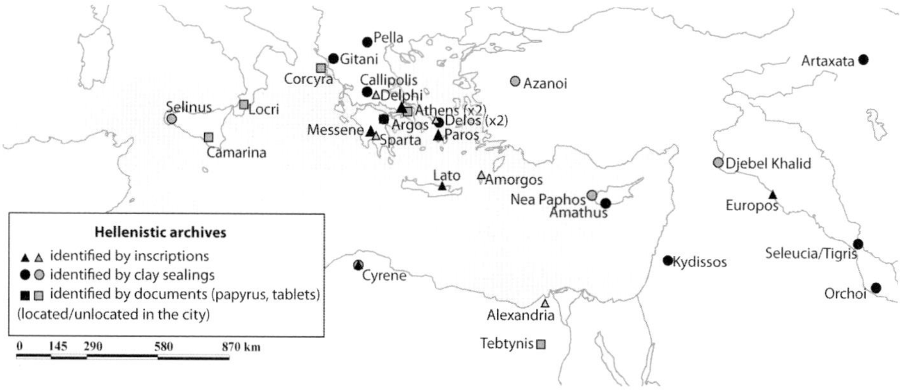

Figure 2.1 Map of known Hellenistic record offices (excluding archives only known from one inscription) [© G. Coqueugniot]

in the houses of merchants or politicians.[7] While frequently mentioned in the resolution of inscribed decrees and in historical accounts, these archives, generally written on perishable material, have often completely disappeared without a trace in the archaeological record. In fact, most of the Hellenistic archival offices we know today[8] were identified either by an associated inscription[9] or by the discovery of clay sealings.[10] These sealings were stamped lumps of clay that were once appended to close the documents. When the documents were destroyed by fire, the sealings were then baked and were thus preserved until their excavation. The buildings themselves generally lack distinctive architectural features: public records were stored in ordinary, quadrangular rooms in larger complexes, either sanctuaries or civic offices, while private records were kept in the individual's house.[11] Like the documents, the furniture often disappeared with only limited indication of their presence and shape apart from nails and burnt fragments of wooden boards. Smaller collections were apparently stored in wooden boxes or occasionally in large vessels, while more important archives were placed on shelving units, such as the mud-brick pigeonhole filing system discovered in the *chreophylakeion* of Europos-Doura.[12]

When studying the place and use of archival records in the Hellenistic world, we are thus confronted by important difficulties resulting from the scarcity of the sources and their elusiveness. First, as mentioned above, few actual records survived

7 See for example the merchant's archive in the 'House of the seals' at Delos (Coqueugniot [2013] 93–96) and the archive of the *strategos* of Callipolis (pp. 80–81).
8 Coqueugniot (2013) 67–150, and *passim.*
9 Christophilopoulou (1979).
10 Boussac/Invernizzi (1996).
11 Coqueugniot (2013) 43–45.
12 Coqueugniot (2013) 48–53. For the archive room at Europos-Doura, see Coqueugniot (2012) 97–98.

Figure 2.2 The archive room in the *chreophylakeion* of Europos-Doura [courtesy Dura-Europos collection, Yale University Art Gallery]

from the Hellenistic period, with the notable exceptions of papyri archives from Lagid Egypt, groups of clay tablets from Seleucid Mesopotamia, and a few collections of lead and bronze tablets from *Magna Graecia*, Sicily, and Greece. Those discoveries, however, all came with other defects that have prevented us from gaining a complete and accurate view of those archives. Indeed, many of these documents came from unrecorded excavations, and we know little to nothing about their context of discovery or even about the coherence of collections often dispersed in the antiquity market.[13] In addition, even when they were found in regular, well-recorded excavations, most of the archival *corpora* we know come from a secondary context, after they had been discarded and removed from their original repository. For example, many outdated Lagid *papyri* were sold to embalmers and reused in mummy cartonnages,[14] while others were thrown in rubbish dumps in the desert just outside of the villages.[15] Cuneiform clay tablets could be reused as building material to fill in walls and benches,[16] and lead tablets were sometimes thrown away in public

13 A large part of the work of modern papyrologists interested in these archives consists in piecing together documents from a same archive—sometimes from a same papyrus roll even—that were sold to different collectors and museums across the world. Vandorpe (2009) 224–26.

14 Salmenkivi (2002) 9–13.

15 For example, many Hellenistic and Roman documents were found in rubbish heaps near the main temple of Tebtynis. Gallazzi (2012) 106–10.

16 Veenhof (1986) 2.

wells.[17] Paradoxically, excavations have mostly covered those records that were discarded in antiquity after their immediate value disappeared. On the other hand, other archival documents whose value outlived their immediate practical use are only attested in indirect, literary or epigraphic *testimonia*. While they probably would have presented a keener interest to ancient historians and other scholars, they generally did not leave any trace in the archaeological record.

Other, significant difficulties lie in the identification of archival records, in opposition to either non-archival documents or literary writings, and the ambiguity of the ancient vocabulary referring to both archival repositories and their contents. Since the late Archaic and Classical periods, there is evidence for a growing concern from ancient Greek cities and individuals to use writing in their day-to-day life and to record important decisions and transactions. The date when cities came to rely more heavily on written documentation and thus developed archival repositories or the habit to publish important decisions in their civic and religious centres has been central to the debate between classicist scholars on the place of literacy versus orality in the ancient Greek world.[18] For a long time, and still today, classicists did not draw a clear distinction between archival documents and inscriptions. Archival records, however, were preserved in a closed repository and only accessible for those allowed to consult its holdings. On the contrary, inscriptions were published versions of texts. Indeed, a text may have had multiple versions inscribed and archived.[19] More than the informative value they have for today's scholars, inscriptions had a significant symbolical function in antiquity, which appears to be largely lacking in archival documents: they served to advertise the inscribed text to the community.[20] The archival version of one text is only seldom mentioned in inscriptions, although some indications show that the practice of preserving copies of most acts in public and private repositories was implicitly expected. A series of Hellenistic inscriptions, however, show the reinforcement of archiving modalities in the Greek cities, whether by the creation of a double archiving or by the transfer of private or semi-private archiving practices to the state.[21]

Moreover, it is sometimes difficult to identify the mention of these archival documents in literary and epigraphic sources, as the vocabulary used to refer to this is

---

17 Several batches of lead tablets from the cavalry archive in Athens were found in two wells in the agora and Kerameikos. Kroll (1977) 84–85.

18 Athens is usually considered as one of the most document-minded Classical cities, which developed a strong epigraphic habit and whose archival repository in the *Metroon* is already attested in the late 5th century BCE, in opposition to Sparta, where inscriptions were rare and of a later date. For an overview on the question of literacy, see Yunis (2003).

19 See the cautioning against this confusion in Robert (1961) 458–59.

20 Boffo (1995).

21 In 2nd century BCE Paros, the archive of the *mnamones* is duplicated and deposited in a second archive in the sanctuary of Hestia: Lambrinudakis/Wörrle (1983). From the late 1st century BCE, Delphian manumission acts, previously archived by private individuals, are to be deposited in a public repository: Mulliez (1992) 34–36.

fluctuant. There is not one unique word to designate archival documents: they are generally referred to by their technical nature 'decree', 'law', 'letter', 'account', etc. rather than by their archival status, although mentions of the material of these documents (wooden tablets or sheets of papyrus or parchment), which imply their conservation in an archival office rather than in an open setting, are increasingly frequent in the Hellenistic period.[22] In addition, the Greek words used to designate these archives often present a polysemy that makes most identification uncertain.[23] Thus, alongside the less ambiguous τὰ ἀρχεῖα or τὰ δημόσια γράμματα (the 'public letters') frequently used to designate the public archives or the building where they were preserved, we find words that can refer to completely different kinds of writings: ἡ ἀναγραφή can thus be indistinctly a public register in the archives,[24] an inscription,[25] or a philosophical treatise[26] while τὰ ὑπομνήματα designate either archival registers[27] or historical memoirs.[28]

Polybius, Timaeus, and Archival Research

In the 2nd century BCE, Polybius wrote a violent criticism of one of his most famous predecessors, Timaeus of Tauromenion, who lived in the late 4th and first half of the 3rd centuries BCE. He condemned most particularly the historical method of the latter, which he considered focused exclusively on sedentary research in the libraries and archives of Athens:

> [2] ἧττω δὲ τῶν ὁδῶν ὥρμησε πρὸς τὸ πολυπραγμονεῖν. [3] τῶν μὲν γὰρ διὰ τῆς ὁράσεως εἰς τέλος ἀπέστη, τῶν δὲ διὰ τῆς ἀκοῆς ἀντεποιήσατο. καὶ ταύτης ⟨δι⟩μερ⟨οῦς⟩ οὔσης τινός, τοῦ μὲν διὰ τῶν ὑπομνημάτων. . . τὸ δὲ περὶ τὰς ἀνακρίσεις ῥᾳθύμως ἀνεστράφη, καθάπερ ἐν τοῖς ἀνώτερον ἡμῖν δεδήλωται. [4] δι' ἣν δ' αἰτίαν ταύτην ἔσχε τὴν αἵρεσιν εὐχερὲς καταμαθεῖν: ὅτι τὰ μὲν ἐκ τῶν βυβλίων δύναται πολυπραγμονεῖσθαι χωρὶς κινδύνου καὶ κακοπαθείας, ἐάν τις αὐτὸ τοῦτο προνοηθῇ μόνον ὥστε λαβεῖν ἢ πόλιν ἔχουσαν ὑπομνημάτων πλῆθος ἢ βυβλιοθήκην που γειτνιῶσαν. [5] λοιπὸν κατακείμενον ἐρευνᾶν δεῖ τὸ ζητούμενον καὶ συγκρίνειν τὰς τῶν προγεγονότων συγγραφέων ἀγνοίας ἄνευ πάσης κακοπαθείας.

22 Scafuro (2013) 402.

23 Georgoudi (1988) 225 discusses the Greek terms indicating an archival process, especially the verb ἀναγράφω that indicates either the registration of a document in the archives or its publication on a stele.

24 Plato, *Leg.* 850a mentions the register of properties held by metics in Athens.

25 For example, in Amorgos' 3rd century BCE decree honouring Hegesippos and Antipatros, *IG* XII 7, 386, l. 43. The corresponding verb is ἀναγράφω, which appears in the instructions for publication of numerous decrees requesting to 'inscribe the decree on a stele': ἀναγρά[ψαι] | [δ]ὲ τόδε τὸ ψήφισμα εἰς στήλην (*IG* XII 7, 386, l. 40–41).

26 Plutarch, *Moralia* 1012b.

27 Plutarch, *Moralia* 867a.

28 Polybius, *Hist.* 1.1.1.

[2] Now, Timaeus enters on his inquiries by the pleasanter of the two roads, but the inferior one. [3] For he entirely avoids employing his eyes and prefers to employ his ears. Now the knowledge derived from hearing being of two sorts, Timaeus diligently pursued the one, the reading of documents, as I have above pointed out, but was very remiss in his use of the other, the interrogation of living witnesses. [4] It is easy enough to perceive what caused him to make this choice. Inquiries from books may be made without any danger or hardship, provided only that one takes care to have access to a town rich in documents or to have a library near at hand. [5] After that one has only to pursue one's researches in perfect repose and compare the accounts of different writers without exposing oneself to any hardship. (Polybius, *Histories* 12.27.2–5)[29]

In this extract, Polybius insists on the primary importance of direct investigation in the elaboration of a historical method: the historian must be a direct witness of the events he relates or, if that was not possible, he has to rely on the testimonies of direct witnesses. Timaeus, who constructed his historical works mostly from careful literate research, without travelling much outside of Athens, is thus presented as the counter example of a good historian. We should, however, not take strictly at face value this criticism of Timaeus, an early Hellenistic historian only preserved in fragments.[30] Timaeus' work focused on the history of the Greek West from the origins to his time. About half of his work thus concerned a period and events he could not have seen directly and for which he could not rely on eye witnesses, but only on previous writings. The aim of his history is also different from Polybius' work. While the latter is mainly concerned with the political and military history of his time, Timaeus' scope appears larger, both in its chronology and the themes he developed. For example, he included stories on the origins of various cities, their customs, and curiosities, in an almost Herodotean conception of history. In addition, several fragments attest that Timaeus, although he wrote most of his works in Athens, also conducted research in other Greek cities; when discussing the creation of the Italian colony of Locri Epizefirii, Timaeus claims he consulted treaties and decrees of the Greek Locrians.[31]

Despite his firm condemnation of Timaeus' method in Book 12, Polybius seems to consider him a worthy predecessor in the introduction and the conclusion of his work.[32] By minimising the importance of literary and documentary research for the writing of history, for which Timaeus was well-known,[33] Polybius aims at enhancing

29 Trans. adapted from Paton (1925).
30 Baron (2013) 58–88.
31 Polybius, *Hist.* 12.7–9.
32 Polybius, *Hist.* 1.5.1; 39.8.4.
33 Polybius, *Hist.* 12.10.4: καίτοι διότι τοῦτ' ἴδιόν ἐστι Τιμαίου καὶ ταύτῃ παρημίλληται τοὺς ἄλλους συγγραφέας καὶ καθόλου τῇδέ πῃ τῆς ἀποδοχῆς . . .—λέγω δὲ κατὰ τὴν ἐν τοῖς χρόνοις καὶ ταῖς ἀναγραφαῖς ἐπίφασιν τῆς ἀκριβείας καὶ τὴν περὶ τοῦτο τὸ μέρος ἐπιμέλειαν—δοκῶ, πάντες γινώσκομεν. ('And yet Timaeus's special boast, the thing in which he outvies other authors and which is the main cause of the reputation he enjoys, is, as I suppose we all know, his display of accuracy in the matter of dates and public records, and the care he devotes to such matters').

his own historical method, which relies more on direct experience. Timaeus is thus deemed a 'lazy historian' spending his time in the libraries of Athens, where he lived most of his life in exile. Examining more precisely the text, however, we can see that Polybius mentions two different types of writings used by Timaeus, ὑπομνήματα (*Hist.* 12.27.3) and βυβλία (12.27.4). As we already discussed above,[34] the term ὑπόμνημα is often difficult to translate, as it may refer to different kinds of writings, either archival registers or literary and historical chronicles. In the following sentence, Polybius poses the preliminary conditions to Timaeus' method: 'to have access to a town rich in documents or to have a library near at hand.' There, the ὑπομνήματα are thus clearly identified as writings that are not to be found in a library, thus indirectly implying the use of archival registers. While archives and libraries were both attached to the conservation of writings, and thus present some similarities in their architecture and organisation, their aim ultimately differed in the nature of the writings held by either, as well as in the associated activities also hosted in or around the buildings.[35] On one hand, archival offices were mainly associated with governmental bodies or commercial centres. On the other hand, libraries are rather identified in cultural or educational centres of the city, such as the *gymnasium*, and in association to structures allowing the gathering of scholars for discussion and debates and the diffusion of literary works.[36] Another fragment also mentions Timaeus' use of ἀναγραφαί of various types: a 'written treaty' (συνθήκας ἐγγράπτους) and 'decrees' (δόγματα).[37] These documents may very well have been archival records the Greek Locrians presented to him during his inquiries.

Despite Polybius' critique of his historical method, Timaeus was a respected and popular historian in the Hellenistic period. He relied mostly on secondary research in the libraries and archives, and spent most of his life in exile in Athens, which was well-equipped with books and documents at that time. Was his approach to history, using archival and non-archival documents alike, exceptional at that time, or can we see it as typical of late Classical and early Hellenistic scholarship in centres like Athens?

---

34 See nn. 27–28.

35 Coqueugniot (2013).

36 See in particular the contribution by Matthew Nicholls on 2nd century CE libraries and their insertion in the literary landscape of central Rome, in this volume.

37 Polybius, *Hist.* 12.9.

Scholars and their Use of Documentary Evidence in Classical and Early Hellenistic Athens

The use of epigraphic and archival documents is limited, but well-attested in the Athenian scholarship of the Classical and early Hellenistic period.[38] Most of the evidence for the growing scholastic use of documents consists, however, in the mention of inscriptions rather than archival records. Thus, Thucydides used two epigraphic dedications erected in Athens to argue that Pisistratus the Younger held the archonship at the end of the Archaic period.[39] While such epigraphic references appear only sporadically in fifth-century literature, their use greatly increased beginning in the second half of the 5th century and especially in the 4th century BCE. This increase thus appears to parallel the growing number of inscriptions during that period. In the 4th century, we can also observe the appearance of archival documents alongside epigraphic writings as sources for several scholars, especially the Athenian orators and the Atthidographers.

Athenian orators are among the scholars whose use of documents is the best attested in antiquity, using inscriptions and archival records as pieces of evidence in their judiciary and political discourses. This use is made very publicly, and is even sometimes emphasised. The orators often mentioned their documentary sources and occasionally specified where one inscription or record could be seen:

> ἀλλ' ὑπὲρ μὲν τῆς ἐξωμοσίας ἐν τοῖς κοινοῖς τοῖς ὑμετέροις γράμμασιν ἐν τῷ μητρῴῳ ταῦτ' ἐστίν, ἐφ' οἷς ὁ δημόσιος τέτακται, καὶ ψήφισμ' ἄντικρυς περὶ τούτου τοῦ ὀνόματος γέγραπται.
>
> As for the affidavit of refusal, there is an entry in your public records in the *Metroon*, of which the public caretaker is in charge, and a decree in which he is mentioned by name. (Demosthenes, *On the false embassy* 129)[40]

The *Metroon*, the sanctuary to the Mother of Gods, is located in the south-west of the Athenian agora, next to the new *bouleuterion* and the *tholos*.[41] It took over the space of the ancient *bouleuterion* destroyed during the Persian wars, and was rebuilt in a grander, monumental scale in the 2nd century BCE. It was used as the official repository for the archives of the *Boule* and the *Ekklesia* from the last decade of the 5th century BCE at the latest. It is there that all decrees and laws of the Athenian democracy were kept, under the care of the 'secretary of the *Boule*' and a public slave. Demosthenes clearly states all Athenian citizens could consult the archive on demand.[42] He and his contemporary Aeschines regularly mention their use of the

38 Rhodes (2007) 56–66.
39 Thucydides, *Hist.* 6.54.6–7.
40 Trans. adapted from Vince (1926)
41 Shear (1995) 157–90.
42 Demosthenes, *Against Aristogeiton* 99. This is also implied in Aeschines, *Against Ctesiphon* 75.

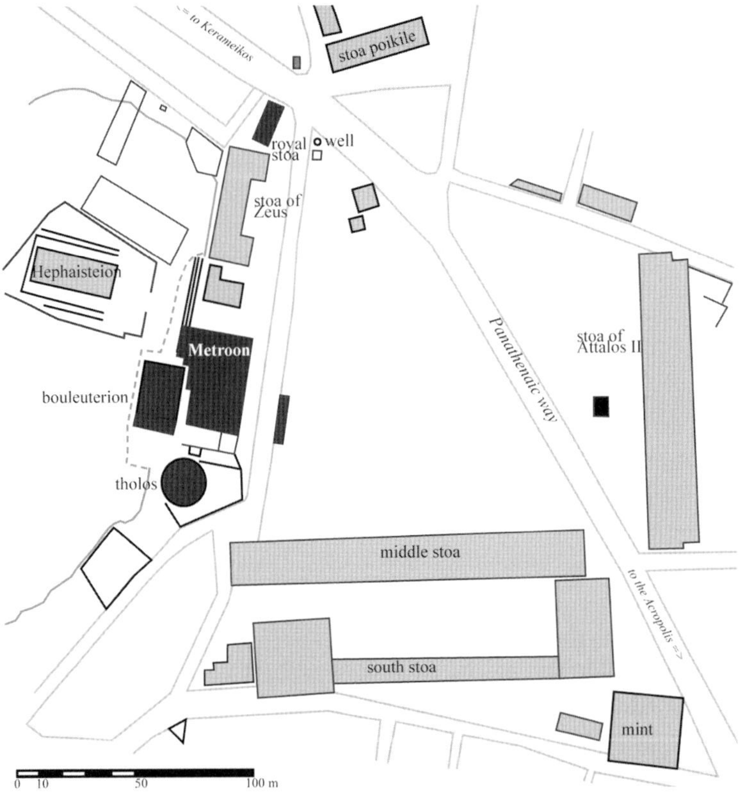

Figure 2.3 The Athenian agora in Hellenistic times [drawing G. Coqueugniot, after J. Travlos]

decrees or other documents preserved in the *Metroon*.[43] Of course, the reference these two orators made to the archives in the *Metroon* and in other public repositories does not actually guarantee that they really consulted these archives,[44] especially if an epigraphic version of the same decree was more readily available in the city.[45] However, it is now largely admitted that not all decrees were inscribed and published on stone or metal steles. Many decrees quoted by Demosthenes in his speeches concern ephemeral matters that were not usually published on stone.[46] He must consequently have consulted an archived copy of these texts. The same use of the *Metroon*'s archives can also be inferred for other 4th century orators, who do not disclose their

43 For example, Aeschines, *On the Embassy* 58–59, and *Against Ctesiphon* 24; Demosthenes, *On the crown* 142. See Sickinger (1999) 160–75.

44 Lane Fox (1994) 140–41.

45 *In contrario*, Georgoudi (1988) 241 thinks that the text from some inscriptions quoted by Athenian orators (for example, Andocides, *On the mysteries* 96) more probably came from the archival version of the decree, which would have been easier to retrieve and copy for the public secretary.

46 Lambert (2016) 220–23.

sources' location. Of course, it could be argued that the use of archival records in the context of law-courts or even that of political oratory does not directly relate to scholarly research *per se*. This possible objection can however not be made against the following users of archival records in Classical and Hellenistic Athens.

The Atthidographers were scholars who wrote histories of Attica, the *Atthides*. The most famous are Hellanicus of Lesbos in the late 5th century BCE, Androtion in the first half of the 4th century, and Philochorus in the early 3rd century, although they are only known through fragments.[47] Among those fragments not concerned with Athens' mythical period, there are a striking number of quotations implying the recourse to lists or other documentary evidence from Archaic or 5th century Athens.[48] Most of the *Atthides* were organised in a chronological order following the eponymous archons' dating system that was also used in Athenian decrees and laws. An inscribed list of archons existed from at least the late 5th century, but it was less detailed than some of the Atthidographers' references. For example, it didn't include their demes, contrary to Philochorus' work.[49] In addition, while the Atthidographers clearly relied on older literary sources such as Herodotus and Thucydides for some episodes, they occasionally added details that very probably came from archived lists.[50] We also know of one instance at least when Androtion used public records outside of his scholastic work: in 347–346 BCE, he proposed a decree granting honours to the rulers of Pantikapaion in the Pontus.[51] This decree directly refers to another honorific decree to their grand-father and father, which Androtion had consulted. Unfortunately, the elusiveness of the Atthidographers' fragments makes any reference to their sources unsure. Their main sources appear to have been oral or literary works, but they also occasionally made use of documentary sources, either inscriptions or archival records, to complete their account. The place of records in the account seems also to rise, from the oldest Atthidographer Hellanicus, in the 5th century, to the latest ones in the Hellenistic period.

Atthidographers were in turn used by late Classical and Hellenistic scholars. Thus, Androtion is considered as one of the main sources for Pseudo-Aristotle's *Athenaion Politeia*. This work is part of a collection of 158 *politeiai* attributed to Aristotle or his school in antiquity. It is the only one well preserved. The first part of the work is mainly based on previous historical accounts by Herodotus, Thucydides and the Atthidographers, although it also makes reference to parts of the laws of Solon that had already fallen into disuse.[52] This could indicate that the author got access to a complete transcription of this Archaic text, either by consulting the original *kyr-*

47 Jacoby (1949), while still the reference, has been widely completed and commented in the last twenty years. See also Harding (2008).
48 Harding (2001) 43–47.
49 Rhodes (2014) 18–20.
50 Rhodes (2014) 24–25.
51 *Syll.*^{3} 206. Sickinger (1999) 181.
52 Rhodes (2004) 76–78.

beis or *axones*,[53] or through another source. The second half of the work describes the working of Athens' constitution in the 4th century. It was the first work on the subject, and 'must have been based on direct research'.[54] That at least part of this research may have taken place in the *Metroon* is probable as this was the place where all the laws and decrees were preserved, and we know from the orators that Athenian citizens could consult them.[55] In addition, the functioning of some magistracies described in the *Athenaion Politeia* was obviously drawn from direct contact with the records they produced. We can even find in this work direct reference to the physical setting of some archival repositories. For example, the *Poletai*'s tablets recording debtors were kept on ἐπιστύλια, a kind of bookcase.[56]

Finally, a third kind of historical research dealing with records in 4th century BCE Athens concerns the compilation of legal and political documents. We do not know much about Theophrastus' *Nomoi*, but Kraterus' *Collection of Decrees* was widely used as a source on Athenian decrees by later Hellenistic and Roman scholars.[57] Written in the late 4th or early 3rd century BCE, these nine books reproduced the complete text of 5th century decrees with a brief commentary.[58] Some of these decrees were copied from inscriptions, and others certainly came from the *Metroon* archive.

Despite the fragmentary state of most historical and oratory works of the late Classical and Hellenistic periods—with the exception of Demosthenes' and Aeschines' discourses—, direct and indirect reference to the use of archival documents in historical and other scholarly works steadily increased from the late 5th century BCE. These references are paralleled by an increase in the epigraphic documentation discovered in Athens, and it is not always clear if the decree mentioned in an account came from an archival copy or from an inscription. There is, however, direct references to the conservation of some texts quoted in the orators' speeches in the *Metroon*, as well as evidence that copies of documents from these archives were accessible to all Athenian citizens upon request. The accessibility of archived decrees and laws is particularly emphasized in the case of judiciary discourses, where the orators requested the direct reading of these documents by the public clerk.[59] The use of archives by the orators corresponds mostly to a professional use aiming at an immediate, practical purpose in the courtroom or the assembly. While copies of various archival acts could be requested from the archives' personnel, evidence for a direct perusing of the documents for historical research is more rarely attested by direct

53 We know that *kyrbeis* and *axones* with these laws were kept from the 5th century BCE in the agora and the prytaneum, where Pausanias could still see them in the 2nd century CE. See Stroud (1979).
54 Rhodes (2004) 83.
55 Sickinger (1999) 182.
56 Aristotle, *Athenaion Politeia* 47.5.
57 Carawan (2007), consulted online on 5 April 2017.
58 Sickinger (1999) 183–84.
59 See for example Aeschines, *Against Ctesiphon* 75 and Demosthenes, *On the false Embassy* 129.

quotations. This does not, however, mean that such consultation is to be excluded, especially in regard to our highly fragmentary knowledge of Classical and Hellenistic Athenian chroniclers. Indeed, the use of the eponymous archons' list for the dating of Athenian events by the Atthidographers has been clearly demonstrated, as well as their mention of details not available in known inscriptions.[60] The increasing use of archival documents is also apparent in the development of encyclopaedic scholastic research in the late 4th century BCE, around the school of Aristotle. Works on the Greek constitutions necessitated the gathering of numerous *testimonia* on the functioning of the states, including laws and decrees that had already fallen out-of-use and would therefore not have been displayed in public. Kraterus' work also aimed at making more easily available legal texts copied from inscriptions or archives. These examples attest of an evolution of mentalities in the late Classical and Hellenistic periods, when some outdated documents began to attract the attention of historians and other scholars.

The Role of Archives in Local Histories of the East

Apart from Athenian scholars, we have testimonies of several other historians who used ancient records to document their historical works. Roman and Byzantine authors especially mention the use of archives made by Oriental historians of the Hellenistic times to write histories of their people before Alexander the Great's conquest. For example, in his late 2nd century CE *Apology of Christianity*, Tertullian stands up for the great antiquity of the scriptures and exhorts his readers to follow the example of his Hellenistic predecessors who used ancient archives to write the history of their people:

> *Reseranda antiquissimarum etiam gentium archiva, Aegyptiorum Chaldaeorum Phoenicum;* [6] *advocandi municipes eorum per quos notitia subministrata est, aliqui Manethon Aegyptius et Berosus Chaldaeus, sed et Hiero(ni)mus Phoenix, Tyri rex, sectatores quoque ipsorum Mendesius Ptolemaeus et Menander Ephesius et Demetrius Phalereus et rex Iuba et Apion et Thallus et si quis istos aut probat aut revincit; Iudaeus Iosephus antiquitatum Iudaicarum vernaculus vindex.*

> We must open the archives of the most ancient nations, the Aegyptians, the Chaldaeans, the Phoenicians; we must likewise summon to our aid the fellow-countrymen of those from whom our knowledge is gained, Manetho the Egyptian, Berossus the Chaldean, Hieronymus (or Hieromus[61]) the Phoenician, king of Tyre; and their disciples Ptolemy of Mendes, Menander of Ephesus, and Demetrius of Phalerum, king Juba, Apion, and Thallus; and their critic, Josephus the Jew, the native defender of Jewish antiquities, who either confirms their accounts or convicts them of error. (Tertullian, *Apology of Christianity* 19.5–6)[62]

---

60 Stroud (1978) especially 32–34.

61 On a discussion about the two possibilities, see Lopez-Ruiz (2008), consulted online on 5 April 2017, commentary to BNJ 787 T1b.

62 Abbreviated in BNJ 783 T 2a=BNJ 787 T1b. Trans. after Bindley (1890).

While most of these ancient authors, with the exception of Josephus, are only preserved in fragments today, there is strong evidence for their consistent use of ancient records from their fatherland, as the three following examples will demonstrate. Manetho and Berossus are certainly two of the most famous Hellenistic historians from the East, who wrote histories of their fatherland in Greek. New studies on the later, especially, have allowed for a better understanding of his position in Hellenistic Babylon, the value of his work, and the sources on which he based his investigation.

Berossus, who lived in Babylon under the first Seleucids in the late 4th and early 3rd centuries BCE, wrote a history of Babylon in three books.[63] His *Babyloniaka*, which were written in Greek and dedicated to Antiochus I, are known from several later writers, especially Josephus, Eusebius in his Armenian chronicles, and the *Chronography* of the Byzantine monk Synkellus in the 9th century CE:

> Βηρωσσὸς δὲ ἐν τῆι πρώτηι τῶν Βαβυλωνιακῶν φησι γενέσθαι μὲν αὐτὸν κατὰ Ἀλέξανδρον τὸν Φιλίππου τὴν ἡλικίαν, ἀναγραφὰς δὲ πολλῶν ἐν Βαβυλῶνι φυλάσσεσθαι μετὰ πολλῆς ἐπιμελείας ἀπὸ ἐτῶν †που ὑπὲρ μυριάδωνιε περιεχούσας χρόνον· περιέχειν δὲ τὰς ἀναγραφὰς ἱστορίας περὶ τοῦ οὐρανοῦ καὶ θαλάσσης καὶ πρωτογονίας καὶ βασιλέων καὶ τῶν κατ' αὐτοὺς πράξεων.
>
> Berossus says in his first book of the *Babyloniaka* that as for his age, he lived in the time of Alexander the son of Philip. Records of many had been preserved with great care in Babylon, which encompassed a period of somewhat more than 150,000 years ago. The records comprised stories about the sky and the sea, the first generations and the kings and the achievements in their reigns. (Synkellos, *Chronography* F1b)[64]

While the timeframe appears grossly overstated today in this account, the mention of the preservation of old Babylonian records (ἀναγραφαί) and their use by Berossus is of particular interest. Our understanding of this historian has been greatly enhanced with the publication of a growing number of cuneiform tablets and inscriptions from the Hellenistic period in the last fifty years: lists of kings and sages, building inscriptions, Babylonian chronicles, literary account, chronographic documents, astronomical diaries, and eventually administrative and economic records, although the oldest ones would already have been discarded.[65] Indeed, the narrative in some of Berossus' fragments shows a striking parallel with recently discovered cuneiform documents, such as the list of antediluvian sages and kings found in Uruk.[66] Additionally, Berossus' account of the end of the reign of king Nabopolassar is very close from that found in Babylonian chronicles,[67] and his description of Nebuchadnezzar's palace follows the text of the royal Basalt Stone Inscription.[68] The few differences from

63 de Breucker (2013) 16–21.
64 = BNJ 680 F 1b, trans. from de Breucker (2010), consulted online on 1 August 2017.
65 Beaulieu (2006) 118 and 121–22.
66 Beaulieu (2006) 118.
67 Beaulieu (2006) 123–125.
68 van der Spek (2008) 296–300.

these accounts seem to result mostly from Berossus' reinterpretation and adaptation of these sources to a Hellenistic context.

As a priest to Bel-Marduk at Babylon, he would have been closely linked to its sanctuary, the Esagil. He would consequently have had access to the old and extensive libraries and archives of this temple,[69] which functioned as a major centre of Babylonian scholarship during the 1st millennium BCE. Questions, however, lie in the exact nature of the collections he accessed in order to document his work: were they archival documents preserved in the temple's archive, or were they considered as literary texts, which were copied and studied in the libraries? The exact finding place of many cuneiform tablets paralleled in Berossus is unfortunately unknown. Moreover, when they were discovered in regular excavations, literary tablets and some documents we may consider archival have not always been found in clearly separated spaces. Purely economic and administrative records, which were without doubt preserved in archives, do not, however, seem to have been used at all by Berossus. They were 'dull records' with an immediate, practical use, which were then discarded when then fell out-of-use: accounts, lists of goods, deeds of ownership and transaction contracts, etc.[70] These records didn't present any historical interest for the ancient historian, and it is only recently that Assyriologists became interested in the information their study could provide. The rare documentary texts that could already present a historical interest in antiquity—such as the royal inscriptions—were in fact recovered in library rooms, together with literary and scientific tablets.[71]

The Egyptian priest Manetho, a contemporary of Berossus, also wrote a history of his land in Greek.[72] This work was extensively quoted by Josephus and later Christian authors who used its chronology as a framework for dating biblical events. In *Against Apion*, Josephus aims at proving the great antiquity of the Jews, using Manetho's account on the Hyksos period (ca. 1700 – 1580 BCE):

> Ἄρξομαι δὴ πρῶτον ἀπο τῶν παρ᾽ Αἰγυπτίοις γραμμάτων. Αὐτὰ μὲν οὖν οὐχ οἷόν τε παρατίθεσθαι τἀκείνων, Μανεθὼς δ᾽ ἦν τὸ γένος Αἰγύπτιος, ἀνὴρ τῆς Ἑλληνικῆς μετεσχηκὼς παιδείας, ὡς δῆλός ἐστιν; Γέγραφεν γὰρ Ἑλλάδι φωνῇ τὴν πάτριον ἱστορίαν ἐκ ⟨δέλτων⟩ ἱερῶν, ὥς φησιν αὐτός, μεταφράσας
>
> I will begin with Egyptian documents. These I cannot indeed set before you in their ancient form; But in Manetho we have a native Egyptian who was manifestly imbued with Greek culture. He wrote in Greek the history of his nation, translated, as he himself tells us, from sacred tablets. (Josephus, *Against Apion* 1.73)[73]

---

69 Clancier (2009).
70 Veenhof (1986) 3.
71 Veenhof (1986) 6.
72 Dillery (1999).
73 Josephus, *Against Apion* 1.73 – 92 = Manetho Fr. 42, trans. Waddell (Loeb).

Several elements must be discussed in this extract. First, Josephus presents Manetho as a native Egyptian who received a Greek education (παιδεία). As a priest, he was a member of the native elite and in contact with the new ruling Macedonian elite. He wrote his history in Greek, using some narrative schemes originating in Greek historiography, especially Herodotus.[74] Most importantly, the sources he used were, according to this testimony, ancient records that had to be translated into Greek. The phrase used in Josephus to designate the records used by Manetho is subject to controversy, due to a textual mistake in our manuscript of Josephus' *Against Apion*. Mss L's ἐκ τε τῶν ἱερῶν has been alternatively corrected as ἐκ δέλτων ἱερῶν, literally 'sacred tablets', in the Loeb edition, after Gutschmid's 1893 proposal, or as τε τῶν ἱερῶν ⟨γραμμάτων⟩, 'sacred writings' in Jacoby's edition.[75] Contrary to Berossus' sources, egyptologists and papyrologists have not (yet) identified any of the sources used by Manetho. Despite the manuscript's ambiguity on the exact nature of the writings used by Manetho, his position as a priest in one of Egypt's prominent sanctuaries would, however, have granted him access to records preserved in the temples, whether archival documents on wooden tablets or papyrus, or stone inscriptions.

The third Oriental civilisation rich in ancient records according to Tertullian is the Phoenician one, especially the city of Tyre. While no example of Phoenician literature or records have survived until today, Levantine archives were often considered a reference in Jewish and Christian apologies of the first centuries CE:

> ἔστι τοίνυν παρὰ Τυρίοις ἀπὸ παμπόλλων ἐτῶν γράμματα δημοσίᾳ γεγραμμένα καὶ πεφυλαγμένα λίαν ἐπιμελῶς περὶ τῶν παρ' αὐτοῖς γενομένων καὶ πρὸς ἀλλήλους πραχθέντων μνήμης ἀξίων.
>
> Very old public records kept and carefully preserved by the Tyrians deal with their history and with memorable deeds done by others. (Josephus, *Against Apion* 1.107)[76]

The archival nature of the Tyrian documents Josephus alludes to is without ambiguity here: they were public archives recording historical facts pertaining to Tyre and other related people or lands. While the figure of Hieronymus mentioned by Tertullian remains obscure and subject to controversy,[77] other names of historians of Tyre and the Levant have been associated with the consultation of ancient records in the Hellenistic period, such as Menander of Ephesus. Despite originating from Asia Minor, this 2nd century BCE historian is mostly known as Josephus' main source on Tyrian history.[78] Like Manetho and Berossus, Menander 'translated the archival

74 Dillery (2013) 42–47.

75 The 11th century Codex Laurentianus, plut. 69 cod. 22 (= mss L) is the archetype of all extant Greek manuscripts of this work. For a summary of the extract's textual difficulty, see Dillery (2013) n. 40.

76 = BNJ 794 F 1c, trans. Lopez-Ruiz (2008).

77 Some scholars propose to equate the historian mentioned here to Hieronymus the Egyptian, a late Hellenistic or early Roman historian 'who compiled Phoenician Antiquities' (Josephus, *Jewish Antiquities* 1.107 = BNJ 787 F1). See Lopez-Ruiz (2008).

78 Naiden (2008), consulted online on 26 April 2017.

literature of the Tyrians from Phoenician into Greek',[79] thus transmitting a precious testimony to later Greek and Latin authors who didn't master the Tyrian script. The records used by Menander are of various types. In *Against Apion*, Josephus alludes to records of a similar kind as those used by Berossus and Manetho, such as king lists and other chronological lists of deeds:

> Ἀλλὰ πρὸς τούτῳ παραθήσομαι καὶ Μένανδρον τὸν Ἐφέσιον. γέγραφεν δὲ οὗτος τὰς ἐφ' ἑκάστου τῶν βασιλέων πράξεις τὰς παρὰ τοῖς Ἕλλησι καὶ βαρβάροις γενομένας ἐκ τῶν παρ' ἑκάστοις ἐπιχωρίων γραμμάτων σπουδάσας τὴν ἱστορίαν μαθεῖν.
>
> And I shall add Menander of Ephesus. This man has written about the deeds done in the reigns of each king, including deeds involving both Greeks and Barbarians. He took pains to learn their history and drew on the native literature of each. (Josephus, *Against Apion* 1.116)[80]

Another extract clearly demonstrates the kind of archival records Menander may have used in his chronicle: Josephus explicitly quotes Menander when describing the war between Tyre and the Assyrian king Shalmaneser V in the second half of the 8th century BCE.[81] The narrative lists military and diplomatic operations as well as the calamities of the war for the population: rise of the prices of goods, and difficulties in water supply. This does indeed seem consistent with the information one can usually find in archival lists of deeds such as those we know from Babylonia and Egypt.

This overview of the use of archival records by Hellenistic historians from Egypt, Mesopotamia, and the Levant shows similarities in the scholarly context in which these writers lived, both in terms of their method and in their reception. These historians were often members of the local, traditional elite of the temples, although this is not true for Menander, who was a foreigner of Greek origin. They all seem to have had access to old archival records and used them to write a historical chronicle of their land. We are not told how they gained access to these records, probably through their sacerdotal functions in sanctuaries acting as state repositories for Berossus and Manetho. More importantly, the archives they consulted were written in local languages, and a significant part of these scholars' work was their translation of native texts into Greek.[82] Josephus in particular insists on this task of translation. This was essential for the later historians who thus gained indirect access to archival documents they would have been unable to use otherwise.[83]

---

79 Josephus, *Jewish Antiquities* 8.144 = BNJ 783 T 3a: Μένανδρος ὁ μεταφράσας ἀπὸ τῆς Φοινίκων διαλέκτου τὰ Τυρίων ἀρχεῖα εἰς τὴν Ἑλληνικὴν φωνήν.

80 Josephus, *Against Apion* 1.18.116 = BNJ 783 T 3c, trans. Naiden (2008).

81 Josephus, *Jewish Antiquities* 9.283–287 = BNJ 783 F 4.

82 For more on the topic of translating texts, especially from a minority perspective, see the chapter by Sean A. Adams in this volume.

83 See above his remark on Egyptian archives, which he cannot present, and the consequent value of Manetho's testimony, note 71.

Another example about the value of Hellenistic and Roman translations into Greek is the place given to Philon of Berytus in later historiography.[84] While no mention is made of his direct use of archival documents, this early 2nd century CE author translated into Greek the Phoenician chronicles of an early compiler called Sanchouniathon, who himself drew his knowledge of civic and religious matters from Phoenician archives:

> ἱστορεῖ δὲ ταῦτα Σαγχουνιάθων, ἀνὴρ παλαίτατος καὶ τῶν Τρωικῶν χρόνων, ὥς φασι, πρεσβύτερος, ὃν καὶ ἐπ' ἀκριβείαι καὶ ἀληθείαι τῆς Φοινικικῆς ἱστορίας ἀποδεχθῆναι μαρτυροῦσιν. Φίλων δὲ τούτου πᾶσαν τὴν συγγραφὴν – ὁ Βύβλιος, οὐχ ὁ Ἑβραῖος – μεταβαλὼν ἀπὸ τῆς Φοινίκων γλώττης ἐπὶ τὴν Ἑλλάδα φωνὴν ἐξέδωκε (...) Σαγχουνιάθων δὲ ὁ κατὰ τὴν Φοινίκων διάλεκτον φιλαλήθως πᾶσαν τὴν παλαιὰν ἱστορίαν ἐκ τῶν κατὰ πόλιν ὑπομνημάτων καὶ τῶν ἐν τοῖς ἱεροῖς ἀναγραφῶν συναγαγὼν δὴ καὶ συγγράψας
>
> Sanchouniathon narrates what follows, a man who lived in the earliest times, even before the Trojan War, as they say. He is regarded as trustworthy on account of both the accuracy and veracity of his Phoenician History. His entire treatise was translated from Phoenician into Greek by Philon (the one from Byblos, not the Jew) (...) Writing in the Phoenician language, Sanchouniathon recounted the whole of ancient history inspired by a love of the truth, gathering his material from records of each city and the temple registers. (Eusebius, *Evangelical preparation* 1.9.20 – 21)[85]

This emphasis on the translating act and the content of the preserved narratives, which described the antediluvian origins of their nation, highlight the cosmopolitan context of Hellenistic scholarship. Foreign histories and wisdoms were then made available to the Graeco-Macedonian court elites. Both Berossus and Manetho dedicated their works to Hellenistic sovereigns, and the act of 'translating' the ancient wisdoms into Greek is a recurring theme also at the heart of Alexandrian scholarship in the first centuries of Lagid rule.[86]

One should however remain careful in drawing firm parallels between the three cultures presented in these histories, as they are only preserved in indirect fragments often coming from the same secondary sources. They are mainly used by Josephus, Eusebius, Tertullian, and later Christian writers to demonstrate the great antiquity of the biblical accounts and of 'Barbarian wisdoms' in comparison to more recent Greek literacy. In addition to focusing only on a very small part of Berossus', Manetho's or Menander's histories, we should therefore not forget that we see their work through a deforming lens created by the apologetic context in which they were used.

If we abstract the context of transmission of these historians of the Hellenistic East, we can in fact recognise historiographical tendencies paralleled in many points with local Greek historians such as the Atthidographers. Their works often seem fun-

---

84 Kaldellis/López Ruiz (2009), consulted online on 4 April 2017.

85 = BNJ 790 F 1, trans. Kaldellis/López Ruiz (2009).

86 As illustrated by the famous episode of the Septuagint's translation of biblical texts at the centre of the *Letter of Aristeas*. See Honigman (2003).

damentally patriotic, focusing on their fatherland since mythical times. Their invoked sources were primarily derived from oral tradition and literary predecessors. The temples of ancient Egypt and Mesopotamia were often renowned centres of knowledge, where the priests-scholars established rich libraries of literary, historical, magical and scientific texts. The use of archival documents also preserved in these temples is also attested, but seems to remain minor in regard to that of other sources accessible in the libraries, such as the kings' lists, royal inscriptions and various chronicles. In addition, while their histories drew upon long 'Barbarian' scholastic traditions, the works of Berossus, Manetho, and the historians of Hellenistic Phoenicia were addressed to a Greek audience. They consequently adopted some of the historiographical and narrative schemes already familiar to the Greek scholars. It is especially clear that both Berossus and Manetho followed some of the narrative patterns developed in the Herodotian historiography. The great impact of Herodotus' *Historia* on both the Atthidographers and the Hellenistic historians of the East is equally as important as their use of archival documents to complement their oral inquiries and literary research.

The use of archival documents by Hellenistic scholars writing local histories—whatever the city or region they originated from—is paralleled by their use of inscriptions and literary works from their predecessors and contemporaries. In fact, it appears closely linked to the growing use of written sources as historical evidence, already apparent in some 5th century historians such as Herodotus. Their search for evidence coming from distant times, which cannot be found through reliable oral testimonies, has sometimes been considered an antiquarian's method and opposed to the supposedly more serious historical method of 'global' historians such as Polybius.

Did Polybius Use Archival Records?

As already stated above, Polybius' attack on Timaeus' historical method has sometimes been explained by the different aims of the two historians. Timaeus' work could be equated to the local histories of the Atthidographers and the historians of newly Hellenised regions. Their histories almost embody an antiquarian drive at times, focusing on the past lives of the cities and nations rather than on their present.[87] This drive is paralleled with the growing number of antiquarian lists and compilations, such as the research of Aristotle's school on the constitutions and the collections of decrees and laws that flourished in the late Classical and Hellenistic period. In this context, the value of ancient documents was drastically raised, prompting the 're-discovery' of forgotten laws and decrees and even the forgery of

87 Bravo (2007).

such texts in order to raise the prestige of a family, a group, or a city.[88] In this context, some authors may have tried to reinforce the value of their account by fabricating archival sources supporting their argument.[89] In the 2nd century BCE, Polybius appears to reject this sensational kind of history and to promote a return to a more immediate kind of history, following the model of Thucydides. The historian, Polybius claims, must have a direct, practical knowledge of the historical narratives he relates, either because he took part in the events, or learned about them through the questioning of witnesses.[90]

Despite this apparently very negative view on documentary historical research, Polybius does occasionally use inscriptions or archival documents to prove his point:

> καὶ ταῦτα τῆς ἐπιστολῆς ἔτι μενούσης ἐν τῷ πρυτανείῳ τῆς ὑπ᾿ αὐτοὺς τοὺς καιροὺς ὑπὸ τοῦ ναυάρχου πεμφθείσης περὶ τούτων τῇ τε βουλῇ καὶ τοῖς πρυτάνεσιν, οὐ ταῖς Ἀντισθένους καὶ Ζήνωνος ἀποφάσεσι συμφωνούσης ἀλλὰ ταῖς ἡμετέραις.
>
> ... and this in spite of the fact that the dispatch sent home by the admiral at the very time to the Rhodian *boule* and *prytaneis*, which is still preserved in the *prytaneum* at Rhodes, does not confirm the pronouncements of Antisthenes and Zeno, but my own. (Polybius, *Histories* 16.15.8)[91]

Polybius cannot have consulted the letter in the *prytaneum* of Rhodes himself, as he was still at Rome at that time, and so he must have relied on the testimony of an intermediary witness to relay this information. The mere fact that he was still using this in order to discard his opponents' argument, however, shows that Polybius' view on using archival records as historical evidence is not definite.[92] This seems to be confirmed by several studies on the vocabulary of the Roman treaties with Carthage and Macedon, which have argued quite convincingly that Polybius must have had access to these documents, either in the public archives of Rome or in the private archives of the Scipios.[93] In fact, Polybius explicitly mentions the bronze tablets recording these treaties, which could be seen in the treasury of the aediles, in the Capitoline.[94]

---

88 Robertson (1976). The question of the authenticity of older decrees mentioned in literary sources or re-inscribed in the 4th century BCE is at the heart of many debates among modern classicists. See for example the recent controversy about the decrees inserted in Andocides' *On the Mysteries*. Canevaro/Harris (2012) consider the decrees were forgeries, while Hansen (2016) thinks they are genuine.
89 Haake (2013) 93–96 gives the example of the forged Athenian honorary decree for Aristotle, quoted by Ibn Abi Usaybi'a after an unnamed Hellenistic source. This is also the case of the supposed classical decree allowing bigamy mentioned by Hieronymus of Rhodes in defence of Socrates (Haacke [2013] 100–102).
90 Rhodes (2007) 64.
91 Trans. Paton (1925).
92 Pedech (1964) 379–80.
93 Pedech (1964) 382–389; Stasse (2009) 249–63. Walbank (1990) 82–84 is more circumspect about the possibility of Polybius consulting Roman public repositories before 150 BCE, but he doesn't contest his probable use of the private archives of the Scipios.
94 Polybius, *Hist.* 3.26.1.

Polybius promotes an analytic history relying as much as possible on his own experience or on direct witnesses' testimonies. He is not, however, completely opposed to the use of literary testimonies such as historical works and politicians' memoirs, and documentary testimonies like inscriptions as sources. In fact, much of his work is based on the chronicles and accounts of his predecessors and contemporary historians. He primarily combined and assessed these previous writings, and drew them together in his larger historical project. He also occasionally used archival sources, either directly, such as texts preserved in public and private Roman archives, or indirectly, like the Rhodian letter. The latter shows that Polybius could occasionally put the testimony of archival documents above that of a fellow historian, even one familiar with the events discussed, such as the Rhodian Zeno. Polybius also denies the validity of the historian Philinus' testimony in light of his own archival research: Philinus' description of an additional treaty between Rome and Carthage is deemed a falsehood, on the basis that Polybius did not find the original record of this text in the aediles' archive with other state treaties.[95]

Archival Research in the Hellenistic Scholastic Culture: The Historians and the Archive

While by no means as document-minded as modern scholarship, historical research in the Hellenistic period seems to have made a growing use of both inscriptions and archival records. The elusiveness of references to archival sources in the historical literature of the Hellenistic period has been made harder by a default of identification in our sources, as Hellenistic scholars did not systematically reference their sources. While literary sources have sometimes been identified through careful meta-textual studies, and uncovered inscriptions can be checked against historical works treating of the same event, archival records have most of the time disappeared without leaving a trace. Nevertheless, local historians from Greece and from the East make regular, direct or implicit references to archival documents to back up their accounts. Despite his claims on the greater value of practical knowledge and witnesses' testimonies over written sources, even Polybius occasionally appeals to the testimony of archival records. The use of archives in Hellenistic historical research seems to be growing in the same way as the importance of writing in Hellenistic cities. It is also paralleled with the development of structured public repositories for the documentation produced by the cities' and royal administrations, and sometimes also for private documents.

95 Pedech (1964) 378; Eckstein (2010). Even proponents to the authenticity of Philinus' account against Polybius' accusations do not question the reality of the latter's research in the Roman archives.

Several factors may however have contributed to limit the consultation and the use of archives by Hellenistic historians and scholars. The first one lies in the primary function of archival collections, at least in antiquity: they were usually composed of documents necessary for the running of the state or that of the household. When their original use expired, many of these no-longer 'current records' were discarded. This immediate, utilitarian function does not allow for regular, long-term consultation. Greek archival documents were often written on papyrus or painted on wooden tablets. They were fragile, and allowing their regular consultation by any individual put them at risk of voluntary or accidental corruption. This risk is actually well demonstrated by the tale of Alkibiades' deliberate destruction of a court case in the *Metroon.*[96] In addition, scholastic research in Classical and Hellenistic Greece relied mostly on previous literary works. The Hellenistic historians, including Timaeus and Polybius, had read the works of their predecessors and contemporary scholars. The bases of many historical works in the Hellenistic period were actually the epics of Homer and the histories of 5th century historians Herodotus and Thucydides, which were considered part of the canon of Greek literature even before the end of the Classical period. Other references are sometimes less obvious to modern classicists. The rarity of precise quotations by historians is not an indication of a lack of sources; most quotes were made from memory, and references were generally indirect, vague or even unacknowledged.

The value of direct testimonies, however, tends to increase from the late 5th century BCE onwards, whether they are oral or written testimonies from direct witnesses, or 'original' documents such as inscriptions and archival records. Among the documentary evidence mentioned by Hellenistic scholars, a majority are either inscriptions or compilations of documents. This has actually been explained by the easier access to these published records. While access to many public archival repositories was possible for citizens who requested it, this access was necessarily restricted because of the fragility of the records. On the contrary, inscriptions were generally erected in prominent locations in the city, and exposed to the view of all. Compilations of laws and decrees allowed for the transmission of archival documents to a wider audience and their indirect use by following scholars, who thus could get around the necessity to request access to the original documents. Emphasis on the indirect use of archives granted by chroniclers and compilers is also highlighted by the Roman and Christian writers quoting the historians of the Hellenistic East. In the case of Egyptian, Babylonian and Tyrian archives, the necessity of intermediaries between the archived texts and most historians is made greater by a more restricted accessibility to the repositories and the necessity for the historian to be literate in the non-Greek languages used in these documents. More than a real difference in practice between the historians of mainland Greece and those of the East in their treatment of archival documents, our fragmentary sources rather show several, com-

96 Athenaeus, *Deipn.* 9.407b-c.

plementary aspects of the use of documentary testimonies in the local and regional histories of the Hellenistic period.

Myrto Hatzimichali

Circulation of Lexica in the Hellenistic and Early Imperial Period

Dissemination of Hellenistic Scholarship

Questions about the ways in which ancient texts were published and circulated are very interesting indeed, but often hard to answer, especially since very different circumstances might apply to different time periods and different types of texts. Wondering about dissemination and circulation raises a basic preliminary question as to what publication consisted in for an ancient text. Van Groningen and Dorandi have warned against thinking of publication in modern terms, i. e., in terms of production of multiple identical copies intended for distribution or even for sale.[1] This 'austere view' has recently been called into question, especially regarding the burgeoning book trade in Rome,[2] but it may still hold for the period before Rome's dominance as a cultural centre. The Greek term for 'publication', *ekdosis* (Latin *editio*), in Hellenistic times had already become a technical term denoting the philological enterprise involved in producing a 'critical edition', implying textual criticism and emendation based on conjecture and/or collation of different manuscripts.[3] The primary sense of *ekdosis*, however, is about the author's or proprietor's act of relinquishing control of his creation by making it known to a wider public.[4] The initiative for any further reproduction rested with the prospective users and books were rarely copied without being pre-ordered by their readers (except perhaps in cases of gifts).[5] Diogenes Laertius provides some indications of what this act of relinquishing control might consist in, at least in the case of philosophical writings: it could involve a form of loan charge, allowing interested parties to read the books for a fee, or quite commonly a public reading.

> ἅπερ Ἀντίγονός φησιν ὁ Καρύστιος ἐν τῷ Περὶ Ζήνωνος νεωστὶ ἐκδοθέντα εἴ τις ἤθελε διαναγνῶναι, μισθὸν ἐτέλει τοῖς κεκτημένοις.

I would like to thank the participants of the 2017 Glasgow conference for their questions and comments, and especially Sean Adams for his meticulous editorial work.

1 Van Groningen (1963); Dorandi (2007) esp. 84–85.
2 See the contribution by Matthew Nicholls in this volume, with further references.
3 On this activity, see most recently Montanari (2015a).
4 The relevant verb, ἐκδίδωμι, was typically used in classical times for giving up one's daughter in marriage or renting out property.
5 Thus Van Groningen (1963) 5; Dorandi (2007) 80.

Myrto Hatzimichali, University of Cambridge

https://doi.org/10.1515/9783110660982-004

As Antigonus of Carystus says in his *Life of Zeno*, when the writings [*of Plato*] were first made available, their possessors charged a certain fee to anyone who wished to consult them. (Diogenes Laertius, 3.66)

καὶ δύο μνᾶς αὐτῷ δίδωμι καὶ τὰ ἐμὰ βιβλία τὰ ἀνεγνωσμένα· τὰ δ' ἀνέκδοτα Καλλίνῳ ὅπως ἐπιμελῶς αὐτὰ ἐκδῷ.

I bequeath him two minas and my writings that have been read out, while the unpublished ones I entrust to Callinus, so that he may make them public with care. (Diogenes Laertius, 5.73, from the will of the Peripatetic Lyco)[6]

Hellenistic scholarship stands out as a peculiar case in this regard, both because of the nature of the works produced (we can hardly imagine long commentaries or lexica being read aloud), and because of the political considerations that gave rise to it: there was a sustained drive by the Ptolemies in Egypt, and other Hellenistic monarchies to a certain extent, to accumulate both books and highly educated personnel, in an effort both to assert themselves as the 'true' heirs of Greek civilisation and to promote Greek culture vis-à-vis the conquered civilisations.[7] These cultural politics, culminating in the establishment of the Museum and Library at Alexandria, resulted in several generations of scholars enjoying a privileged status of patronage at court. This, in turn, raises questions about whether the works they produced had any dissemination and impact in the world outside the 'ivory towers' of the great libraries and indeed if there was any interest in them beyond technical scholarly debates. As we shall see in the course of this chapter, there are cases where this can be answered in the affirmative, with important implications for the way in which we understand the publication aspect of ancient scholarship. The earliest such case is Philitas' lexicographical work, which was known at least by reputation to authors of Athenian comedy in the turn of the fourth and third centuries BCE:[8]

ἔθυεν, ἔλεγεν ἄλλα ῥήματα τοιαῦθ' ἅ, μὰ τὴν Γῆν, οὐδὲ εἷς ἤκουσεν ἄν, | μίστυλλα, μοίρας, δίπτυχ', ὀβελούς· ὥστε με τῶν τοῦ Φιλίτα λαμβάνοντα βυβλίων σκοπεῖν ἕκαστα τί δύναται τῶν ῥημάτων.

He made the sacrifice and used countless other words of a sort no one, by Earth, could have understood: *mistulla, moires, diptucha, obeloi*. The result was that I would've had to get Philitas'

6 Another interesting expression for this act of releasing books into the public domain is εἰς μέσον θεῖναι (lit. 'place in the middle'), used by Plutarch (*Sull.* 26) to describe what Andronicus of Rhodes did with the antique copies of Aristotle's works that he received from the grammarian Tyrannio. On Andronicus' publication-related activities, see Hatzimichali (2016).

7 An important source for these developments is the *Letter of Aristeas* (see the contribution by Sean A. Adams in this volume). On Ptolemaic cultural politics, see also Erskine (1995); El-Abbadi (2004); Thompson (2008); Hatzimichali (2013a) 64–68.

8 Philitas' name also appears (as a generally accepted emendation) at *Anth. Pal.* 11.218, an epigram by Crates where the poet Euphorion is mocked for sexual indecency and said to have made full use of Philitas' lexicographical work. For the influence of this work, see Spanoudakis (2002), 396–400.

books to figure out what all the vocabulary he used meant. (Strato, Fr. 1 K-A, transl. Olson = Athenaeus, *Dein.* 9.383a-b)

It is not clear, however, how representative this is for the products of Hellenistic scholarship as a whole. I will attempt to approach the problem in more detail by examining a specific case, that of the lexicographical tradition that was built up during the Hellenistic period primarily through the collection of regional glosses and augmented from other sources, culminating in the monumental work of Pamphilus in the first century CE. This tradition is a partial representation of ancient lexicographical activity, but I believe that it possesses a degree of unity and cohesion that makes it suitable for present purposes. My main source for actual fragments from this tradition will be Athenaeus' *Deipnosophistae*, from the second century CE. The particular value of Athenaeus lies in the fact that, although he has numerically fewer glosses to offer than, for instance, the lexicon of Hesychius or other surviving Byzantine products, his quotations are more extensive, including titles and other details that allow us to form a better picture of the contents and character of the early lexica.[9] This valuable indirect testimony will then be combined with papyrological evidence, which will be a more reliable guide for the access to lexicographical material outside the great libraries, especially since the surviving papyri do not come from Alexandria but from the provinces of the Egyptian hinterland.

The Lexicographical Tradition

Interest in rare words and their meanings was long-standing among the Greeks, even before the establishment of the Museum and the organised study of the literary past. Aristotle defined a γλῶσσα (a rare or exotic word) as a word used by 'others' and clarified that it is a relative term: 'obviously, then, the same word can be both an exotic form and a standard term, though not for the same people: *sigunon* (spear) is standard for Cypriots, an exotic form for us' (*Poetics* 21, 1457b 4–6). Resources for the explication of exotic terms became necessary first and foremost because of the central role played by Homer in Greek education. Homer's text was replete with unfamiliar words that were not part of the vernacular; in fact, the joke in the fragment from Strato's comedy quoted above, and the entire premise of the comic scene that surrounds it, rests on the speaker being unable to understand his cook's Homeric diction. A well-known fragment from Aristophanes' lost play Δαιταλῆς (*Banqueters*, Fr. 233 K-A) places the learning of *glossai* in the context of pupils' daily educational travails.

9 I will use 'lexicon' throughout as a generic term for all these products of scholarship that were structured essentially as word lists. The use of the term here is not meant to imply anything about each work's nature or arrangement. For a discussion of the nomenclature, see Schironi (2009), 3 n.3, and the critique in Esposito (2011) 526.

The Aristophanic fragment is quoted within a text that is particularly representative of the more specialised professional reasons for the development of lexicography, especially in the case of medicine. The source in question is Galen's *Hippocratic Glossary*, which speaks of the need to clarify 'the words that used to be current in the old days, but are not (used) any more nowadays; such words are called *glottai* and these are the ones I am going to explain here' (*Gloss.* XIX 63.1–3 Kühn).

With the growth and increased sophistication of Hellenistic scholarship, interest in *glossai* moved beyond these practical educational and professional purposes, in line with the encyclopaedist tendencies of the Ptolemaic Museum. The minutiae of local knowledge were highly valued by scholars, be it in terms of religious traditions, customs, mythographical details, as well as dialectal variants that were often closely associated with these myths and traditions and were becoming increasingly rare with the spread of the *koine* language. In the very selective examples of Hellenistic glossography quoted below we may observe some of the characteristics of this type of scholarly endeavour. They include, first of all, extensive coverage of different parts of the Greek world, often in terms of highly localised idioms from specific towns (here we have Lesbos, Attica, Eretria, Cyprus, but elsewhere we also find usages e.g., from Corinth, Paphos, Byzantium). We also know of several works with specialised titles like Theodorus' *Attic Glossary* and Diodorus' Italic one that appear in the extracts below. For example, there is a *Laconic Glossary* by Aristophanes of Byzantium (Athenaeus, *Deipn.* 3.77a), a *Rhodian Glossary* by Moschus (*Deipn.* 11.485e) and a *Cretan Glossary* by Hermonax (e.g., *Deipn.* 3.76e).

Φιλητᾶς δ' ἐν τοῖς Ἀτάκτοις ὑποθυμίδα Λεσβίους φησὶν καλεῖν μυρσίνης κλῶνα, περὶ ὃν πλέκειν ἴα καὶ ἄλλα ἄνθη. καὶ ὑπογλωττὶς δὲ στεφάνου ἐστὶν εἶδος. Θεόδωρος δ' ἐν Ἀττικαῖς Γλώσσαις στεφάνων πλοκῆς γένος παρὰ Πλάτωνι ἐν Διὶ Κακουμένῳ.

Philitas in his *Miscellany* (Philit. Fr. 14 Dettori = Fr. 42 Spanoudakis) claims that the Lesbians refer to a twig of myrtle that has violets and other flowers wrapped around it as a *hupothumis*. A *hupoglottis* is also a type of garland; Theodorus in the *Attic Glossary* (*FGrH* 346 F 3b) (says that) Plato in *Zeus Abused* (Fr. 51.2 K-A) uses the word to refer to a style of weaving garlands. (Athenaeus, *Deipn.* 15.678d, transl. Olson)[10]

Διονύσιος δ' ὁ ἐπικαλούμενος Ἴαμβος ἐν τῷ περὶ διαλέκτων γράφει οὕτως· "ἀκηκόαμεν γοῦν ἁλιέως Ἐρετρικοῦ [τὸν] ἱερὸν ἰχθὺν καὶ ἄλλων πολλῶν ἁλιέων καλούντων τὸν πομπίλον· ἐστὶν πελάγιος καὶ παρὰ τὰς ναῦς πυκνὰ φαίνεται ἐοικὼς πηλαμύδι, ποικίλος." [...] Κλείταρχος δ' ἐν ἑβδόμῃ Γλωσσῶν "οἱ ναυτικοί, φησίν, πομπίλον ἱερὸν ἰχθὺν προσαγορεύουσι διὰ τὸ ἐκ πελάγους προπέμπειν τὰς ναῦς ἕως εἰς λιμένα· διὸ καὶ πομπίλον καλεῖσθαι, χρύσοφρυν ὄντα."

The Dionysius nicknamed Iambus writes as follows in his *On Dialects:* 'for example, I have heard an Eretrian fisherman, as well as many other fishermen, refer to the pilot-fish as a sacred fish; it is a sea-fish that often appears alongside their ships, and resembles an immature tuna and is speckled' [...] Cleitarchus says in Book VII of the *Glossary:* 'sailors refer to the pilot-fish as a sa-

10 Translations from Athenaeus throughout are taken from Olson's Loeb edition, unless otherwise indicated.

cred fish because it escorts (*propempein*) their ships from the sea into harbor, which is why it is called a *pompilos*, although it is actually a gilthead'. (Athenaeus, *Deipn.* 7.284b-d)

The reference in this passage to Dionysius Iambus (who was the teacher of Aristophanes of Byzantium, and is therefore dated in the late third to early second century BCE) is of particular interest for the types of sources used by Hellenistic glossographers: they included the spoken language and what is presented as the vernacular of special groups of native speakers like fishermen and sailors. Vocabulary drawn from literary authors had a strong presence in Hellenistic glossography too; Athenaeus' testimony is not as helpful here, because it is not always as clear as in the case of Theodorus' reference to Plato whether a literary quotation comes from a lexicon or not. On the other hand, there is substantial evidence, including papyri, even for lexica devoted to specific authors or literary genres.[11] On the whole, the balance between spoken and literary vocabulary seems to have varied from lexicon to lexicon, and no dominant trend can be detected across all of Hellenistic lexicography.[12]

Another prominent feature in the surviving evidence from Hellenistic lexica is the interest in myths as well as in antiquarian details about the origins of rituals taking place in different cities. Such obscure traditions were understandably the source of rare vocabulary.

Νίκανδρος δ' ὁ Κολοφώνιος ἐν β' Γλωσσῶν βρένθιν λέγεσθαί φησι παρὰ Κυπρίοις θρίδακα, οὗ ὁ Ἄδωνις καταφυγὼν ὑπὸ τοῦ κάπρου διεφθάρη.

Nicander of Colophon in Book II of the *Glossary* (fr. 120 Schneider) says that the Cyprians use the word *brenthis* for lettuce; it was lettuce that Adonis took refuge in when the boar killed him. (Athenaeus, *Deipn.* 2.69b)

Other examples from Athenaeus include Hermonax reporting a ritual related to homosexual love in Gortyn in Crete during the discussion of the word χόννοι (a type of drinking cup) in his Γλῶτται Κρητικαί (*Deipn.* 11.502b); at *Deipn.* 6.235b (= *FGrH* 362.7), Crates (of Athens?) is quoted at length with the exact citation ἐν δευτέρῳ Ἀττικῆς Διαλέκτου ('in the second book of the *Attic Dialect*') on the duties of the ancient officials in charge of grain offerings to the gods for whom the word παράσιτος ('parasite') was originally used.[13] Finally, the scholarly nature of these lexica, which is par-

---

11 The only surviving Homeric lexicon is the one by Apollonius Sophista; there are also surviving Hippocratic lexica by Erotianus and by Galen, as well as a Platonic one by Timaeus the Sophist – all these are dated in the first or second centuries CE. In terms of genre-specific lexica, works devoted to the ten orators are of particular importance (the one by Harpocration, also from the second century CE survives in abridged form), also in the papyrological evidence (cf. *P.Oxy.* XV 1804, 3rd c. CE). Papyri devoted to the explication of words from specific literary genres include *P.Berol.* inv. 9965 (3rd-2nd BCE, Homeric, epic, tragic); *P.Oxy.* XV 1801 (2nd-3rd CE, comic); *P.Sorb.* I 7 (2nd-3rd CE, comic).

12 On these issues, see Ucciardello (2012), 23–24 with n. 31.

13 Broggiato has proved that the identification of the famous scholar from Mallos with this Crates cannot be excluded. On the other hand, the religious / antiquarian nature of the information Atheticularly pronounced in the examples from Pamphilus in the early imperial period to be cited below, is evident from the presence of quotations within the lemmata, both from literature and from earlier lexicographic predecessors and specialist authors.

Modern scholars[14] have also drawn attention to important distinctions between the type of lexica that we have been examining so far, which aimed at gathering and recording regional vocabulary, and others that were author- or genre-specific (e. g., oratory or comedy), and still others that dealt with a specific field of knowledge, such as the *Cookery Terminology* (Ὀψαρτυτικαὶ Γλῶσσαι) by Artemidorus the pupil of Aristophanes of Byzantium (Ἀρτεμίδωρος ὁ Ἀριστοφάνειος, Athenaeus, *Deipn.* 9.387d). The purpose of the thematically specialised lexica is fairly straightforward, namely to aid the reading of the relevant texts, from epic poems to cookery manuals. The existence of such lexica is also testament to increasingly well-defined areas of expertise, developing their 'tools of the trade' and establishing themselves as legitimate areas of study, with their own norms and terminology.[15]

Equally, lexicographical works could prove useful in the process of composition of new literary works, whether one was after rare dialectal or old epic words for their learned Hellenistic poetry, or, in a later period, after well-attested and correct Attic vocabulary for the composition of rhetorical speeches.[16] For this sort of purpose, works organised thematically, known as Onomastica, were doubtless more suitable than alphabetised lists. We know that Callimachus and Aristophanes of Byzantium produced such Onomastica, but the fact is that we have no information as to the arrangement of most Hellenistic lexica.[17]

By comparison, it is somewhat harder to determine any practical purpose for the collections of regional *glossai* like the ones discussed just now, and my tendency would be to fall back on explanations based on encyclopaedic interest emanating from the cultural politics of Hellenistic libraries, especially since this interest is not limited to the linguistic level, but extends to the *realia* behind the words. In

naeus draws from the *Attic Dialect* (cf. also *Deipn.* 11.495a-b = *FGrH* 362.8) could be an argument in support of the identification with Crates of Athens, author of a *On Sacrifices at Athens* (cf. *Suda* EI 184, K 2706 and Eust. *Ad Il.*1283,7 on 22,495, Phot. K 1210). Cf. Broggiato (2000) 364–371.

14 Tosi (2015) 621–36; Matthaios (2010) 167.

15 For parallel developments in how linguistic expertise was conveyed in works of ancient scholarship, see the contribution by Eleanor Dickey in this volume.

16 See Matthaios (2010) 168–69.

17 For some proposals, see Latte (1925) 166–67, 169, 173. The arrangement of many ancient lexica for which we have only indirectly transmitted fragments remains a contested point, see, for example, Tosi (2015) 623–25, whereas lexica preserved on papyrus display alphabetisation (albeit often by first letter only) from as early as the third century BCE (*PHib.* II 175). From the point of view of historical development, I still find compelling the argument made by Alpers (Alpers [1975] 116–17) to the effect that thematic arrangement was a more sophisticated development instigated by Callimachus to complement the pre-existing more mechanical process of alphabetisation. However, see most recently Esposito (2009) 259–60 with further references. We also know of lexica dedicated to specific authors that were arranged according to the order of appearance of the words in the text, see Tosi (2015) 636.

the light of this, we should imagine a limited dissemination, with the depositing of a single copy in the relevant library (not all the lexicographers mentioned so far were necessarily active in Alexandria, for example Nicander of Colophon was associated with the court of Attalus, probably the first). The users of these lexica were for the most part composers of further lexica, in a continuous line of tradition down to Byzantine times. In the 'real world' of the papyri outside the main libraries, the early Hellenistic period (third-second century BCE) offers only brief, elementary word lists most probably from school contexts, ordered alphabetically and concentrating on poetic, mainly epic, words.[18] These include two early texts from El-Hiba (*P.Hib.* II 175, *P.Heid.* I 200), but there is also a Berlin papyrus (third-second century BCE) from the necropolis of Abusir el-Melek (*P.Berol.* inv. 9965) which, despite the brevity of its entries, has been shown by Ucciardello to contain signs of Alexandrian erudition. Moreover, many of its glosses are paralleled in the Lexicon of Hesychius, seven centuries later.[19] The state of the book roll and the text indicate that this was a copy made for private use, perhaps by excerption and compilation from more than one source.[20] This suggests some possibilities for the different stages of publication and dissemination of lexicographical works, whereby they could reach a wider public outside the confines of the Museum. We do not know much about the circumstances in this early period (for more information on the Roman period see below), but we can imagine a scenario where an educated reader visited a central library, perhaps even the royal library of Alexandria (where multiple scholarly resources could be found), copied a selection from different books and brought the result of his labours back to his home town. As for the uses to which a book like *P.Berol.* inv. 9965 could be put, we may think of it as an aid for private reading, but also perhaps as a teaching tool to support the teacher in the explication of challenging vocabulary from Homer and other authors. As Cribiore points out, 'even though some of the glossographical material had an ancient and scholarly origin, the variety shown by the texts of *scholia minora* reveals that teachers felt free to modify, adapt and innovate'.[21]

Post-Hellenistic Repositories of Knowledge

As I have already mentioned, the main consumers of Hellenistic scholarship, including lexicography, were other scholars, just as nowadays scholarly publication is pri-

18 The physical layout of these texts, as far as we can tell from the brief fragments from this period as well from the Imperial period, appears to have been less standardised than the legal texts discussed by Ammirati in this volume.

19 Ucciardello (2012) 21–22. See also Esposito (2009) 287–88.

20 Ucciardello (2012) 34.

21 Cribiore (2001), 207; the *scholia minora* e.g., on Homer are lists of words or short expressions taken from the two epics, 'translated' into an easier form of Greek.

marily aimed at a limited group of like-minded individuals. This phenomenon became particularly pronounced from the Augustan period onwards, when there was a strong tendency towards stock-taking and what I have called 'meta-scholarship',[22] whereby scholars wrote about the work of their predecessors rather than their subject matter directly. The primary exponents were Didymus and Aristonicus, who produced accounts of Aristarchus' edition of Homer and its critical signs, but there are other examples, such as the *On grammarians* by Asclepiades of Myrlea,[23] presenting the history of the subject via its famous exponents. This biographical/historical stock-taking in the first century BCE was paralleled in other subjects such as medicine (Heraclides of Tarentum wrote *On the Empiricist sect*, Gal. *Lib. Prop.* XIX 38; Apollonius Mys and Heraclides of Eythrae both wrote *On the Herophilean Sect*, Str. 14.1.34) and philosophy (the doxographical works attributed to Aëtius and (Arius?) Didymus are thought to originate from this period).

One key characteristic of the works produced from the Augustan period onwards by means of incorporating almost the entire Hellenistic tradition was their length: Didymus 'of brazen guts' (*Suda* Δ 872) is said to have written 3,500 volumes and earned the nickname 'book-forgetter' as a result (Athenaeus, *Deipn.* 4.139c). We also have a reference to the 108th book from the lexicographical work by Dorotheus of Ascalon, who is probably to be dated in the early imperial period:

> Δωρόθεος δ' ὁ Ἀσκαλωνίτης ἐν τῷ ὀγδόῳ πρὸς τοῖς ἑκατὸν τῆς λέξεων συναγωγῆς θέτταν γράφει, ἤτοι ἡμαρτημένῳ περιτυχὼν τῷ δράματι ἢ διὰ τὸ ἄηθες τοῦ ὀνόματος αὐτὸς διορθώσας ἐξήνεγκεν.
>
> Dorotheus of Ascalon in Book CVIII of his *Collected Vocabulary* writes *thetta* (sc. instead of *thraitta*), either because he ran into a defective copy of the play (sc. of the *Horse-groom* by Mnesimachus) or because he introduced the reading himself as a correction, since the name is rare. (Athenaeus, *Deipn.* 7.329d)

To explain this kind of size we have to imagine that Dorotheus' lexicon included a large number of entries, but also that each entry was of considerable length, with pages upon pages devoted to each word, aiming at an exhaustive treatment. In fact, Dorotheus is reported by Porphyry in his *Homeric Questions* to have devoted an entire book to one single entry, on the Homeric word κλίσιον ('shed', 'out-house').[24]

Pamphilus' *On names and rare words* (Περὶ Ὀνομάτων καὶ Γλωσσῶν),[25] produced in the second half of the first century CE, falls into the same category; the *Suda* re-

22 Hatzimichali (2013b) 177–81.

23 Cited in the anonymous *Vita Arati* 1; cf. *Suda* O 657; Π 1888.

24 *Ad Od.* 24.208: ὅλου βιβλίου ἐδέησε Δωροθέῳ τῷ Ἀσκαλωνίτῃ εἰς ἐξήγησιν τοῦ παρ' Ὁμήρῳ κλισίου.

25 I have collected and discussed all the surviving information on Pamphilus in Hatzimichali (2006). The fullest version of the title of his lexicon is given only once by Athenaeus (*Deipn.* 14.650e, cited above), while the *Suda* reports the title *On rare words i.e. vocabulary*, see below.

ports that it was ninety-five books long. Our main source for Pamphilus, just as for an extensive range of Hellenistic erudition, especially in the field of lexicography, is again Athenaeus' *Deipnosophistae*, where Pamphilus' name appears thirty-seven times. These fragments give insight into the kind of content that served to fill up so many books, and the quotations that follow are an indicative sample of the dialects covered, the sources used (both scholarly and literary), and the types of information that were included.

Πάμφιλος δ' ἐν τοῖς περὶ Γλωσσῶν καὶ Ὀνομάτων "ἐπιμηλίς," φησίν, "ἀπίου γένος."

Pamphilus in his *On rare words and names* says: '*epimelis* is a type of pear'. (Athenaeus, *Deipn.* 14.650e, cf. 3.82d)

Διόδωρος δ' ἐν Ἰταλικαῖς Γλώσσαις καὶ Ἡράκλειτος,[26] ὥς φησι Πάμφιλος, τὴν κοτύλην καλεῖσθαι καὶ ἡμίναν, παρατιθέμενος Ἐπιχάρμου· "καὶ πιεῖν ὕδωρ διπλάσιον χλιαρόν, ἡμίνας δύο". Καὶ Σώφρων· "κατάστρεψον, τέκνον, τὰν ἡμίναν."

Diodorus in his *Italic Glossai* and Heracleitus, as Pamphilus reports, say that the *kotyle* [*a volume measurement for liquids*] is also called *hemina*, citing Epicharmus: 'and to drink twice as much warm water, two *heminai*'; and Sophron: 'turn the *hemina*, son, bottoms up!' (Athenaeus, *Deipn.* 11.479a)

καὶ πέμμα δέ τι τευθίδα ὀνομάζειν Ἱατροκλέα ἐν Ἀρτοποιικῷ φησι Πάμφιλος.

Pamphilus says that Hiatrocles in his *Bread-making Book* names a type of cake *teuthis* (squid). (Athenaeus, *Deipn.* 7.326e)

Θεόδωρος δ' ἐν ταῖς Ἀττικαῖς Φωναῖς, ὥς φησι Πάμφιλος ἐν τοῖς Περὶ ὀνομάτων, πλοκῆς στεφάνων γένος τι τὴν ὑπογλωττίδα ἀποδίδωσιν.

Theodorus in his *Attic Words*, as Pamphilus reports in his *On names*, interprets *hypoglottis* as a style of weaving garlands. (Athenaeus, *Deipn.* 15.677b)

Modern scholars have emphasised the fact that Pamphilus' work was the consummate 'general' lexicon, transcending literary, regional, scientific and technical specialisations to encompass all of the Greek language.[27] This doubtless contributed to the length, but it is also clear that Pamphilus' lexicon did not consist simply of words listed with brief definitions, however varied and all-inclusive the vocabulary might have been.

Athenaeus' citations confirm the role of Pamphilus' lexicon as a repository of earlier erudition and of Pamphilus himself as a prime 'consumer' of Hellenistic lexicography. He must have consulted the works in what was for many of them their primary locus of publication, the Alexandrian library itself. We have no way of knowing

26 Ἡράκλειτος cod.: Ἡρακλείδης Wilamowitz: Ἡρακλέων Valckenaer
27 See the lucid analysis of Matthaios (2010) 167–69; also, Ucciardello (2012) 13.

whether Pamphilus travelled further afield (to Rome, for instance), even though we cannot exclude it. Our scant biographical information associates him with Alexandria only, and with the Aristarchean tradition in particular (cf. *Suda* Π 142, cited below). We can be fairly certain, though, that the Alexandrian library in the first century CE would have possessed all the important scholarly works, regardless of where they were initially produced.[28] On seven occasions Athenaeus explicitly states that he has found the reference to some other lexicographer or specialist authority on culinary matters in Pamphilus (the focus on culinary matters is a result of Athenaeus' own interests, and should not be attributed to Pamphilus, other than to surmise that he must have used and quoted similar specialists from other fields of expertise): the references to Diodorus, Iatrocles and Theodorus appear in the extracts cited above.[29] Of course Pamphilus' references to earlier scholarship and reports of his predecessors' views, collected from the large and productive lexicographical tradition of the Hellenistic and Augustan periods, must have extended well beyond the authors mentioned in Athenaeus' extracts. *On names and rare words* also included quotations of literary passages, because references to Hipponax, Epicharmus and Sophron (for the latter see above) as well as to the less famous comic writers Blaesus and Nicon are explicitly ascribed to Pamphilus (see Athenaeus, *Deipn.* 2.69d; 11.487c): in all likelihood there were many more. Thus, the ninety-five books were filled not only with lists of rare and difficult words and their meanings, but also with full citations of the passages where each word occurred (if it came from a literary source) and with reports from earlier authorities. The other type of information that contributed to the length of Pamphilus' work was antiquarian lore on customs, rituals and objects, the collection of which had been an integral part of Hellenistic scholarship ever since Callimachus and was sought and reported by early lexicographers as we saw above. Pamphilus reported on fruit offerings to the gods in Sparta (*Deipn.* 3.82e), on a ceremony performed by young men before the ritual cutting of their hair (11.494f), on the Eleusinian Mysteries (11.486a), and on a sanctuary of 'Aphrodite the prostitute' (Πόρνη Ἀφροδίτη) at Abydos (13.572e).

The publication of Pamphilus' lexicon, like the other scholarly works discussed so far, must have consisted in it taking its place among the holdings of the Alexandrian library. Any further reproduction would have required substantial resources and expense, so it is not possible to say with any certainty whether it ever circulated beyond the library. The only authors for whom we know for certain that they used Pamphilus directly are Athenaeus (a native of Naucratis who could have spent time at Alexandria before moving to Rome and the entourage of his patron Larensis) and Julius Vestinus, who was in charge of the Alexandrian library under Hadrian (to

---

28 See Hatzimichali (2013b) 167–72 for the state of the library's holdings after 48 BCE, with further references.

29 See also *Deipn.* 3.77a for Pamphilus citing Aristophanes of Byzantium; 9.387d for Epainetus, another culinary author, *Deipn.* 11.487b and 14.642e for the lexicographers Apollodorus of Cyrene and Apion.

be discussed below). Even within the confines of the great library, one wonders how such cumbersome works as the lexica of Dorotheus and Pamphilus were intended to be used. Much depends on their arrangement which is often hard to ascertain, but Athenaeus for instance seems to have used Pamphilus not as a reference work for looking up unknown words that he came across elsewhere, but as the primary source for words, definitions and quotations alike. Pamphilus' encyclopaedic ambitions extended not only to the comprehensive coverage of the Greek language, but also to a distillation of previous scholarship and technical writing, and therefore he could have intended his work to be read through as a monumental repository of knowledge about the entirety of Greek culture, and not a reference work, which would require convenience of use and access.

Alternative Modes of Publication and the Fate of Pamphilus' Lexicon

Such monumental ambitions, however, were not compatible with widespread publication and circulation, for practical and material purposes. This is why the fate of Pamphilus' lexicon is a story of successive epitomes and abridgements, with much of its material eventually finding its way into the surviving lexicographical sources from late Antiquity and Byzantium.[30] For these developments we are primarily reliant on the relevant entries of the *Suda:*

> Πάμφιλος, Ἀλεξανδρεύς, γραμματικὸς Ἀριστάρχειος. ἔγραψε Λειμῶνα· ἔστι δὲ ποικίλων περιοχή, Περὶ γλωσσῶν ἤτοι λέξεων βιβλία ϙε′. ἔστι δὲ ἀπὸ τοῦ ε στοιχείου ἕως τοῦ ω· τὰ γὰρ ἀπὸ τοῦ α μέχρι τοῦ δ Ζωπυρίων ἐπεποιήκει. εἰς τὰ Νικάνδρου ἀνεξήγητα καὶ τὰ καλούμενα Ὀπικά, Τέχνην κριτικήν, καὶ ἄλλα πλεῖστα γραμματικά.
>
> Pamphilus, of Alexandria. A grammarian of the school of Aristarchus. He wrote a *Meadow,* which is a summary of miscellaneous contents; *On rare words i. e. vocabulary* in 95 books (it contains entries from epsilon to omega, because Zopyrion had done the letters from alpha to delta); *On unexplained matters in Nicander* and the so-called Opica; *Art of Criticism,* and a large number of other grammatical works. (*Suda* Π 142)

First of all, the entry on Pamphilus includes the claim that his work was not a complete dictionary, but rather a continuation/completion of another scholar's work. The implication is that Pamphilus only worked on words beginning with the letters Ε–Ω, picking up where the otherwise unknown Zopyrion had left off. Regardless of any identifications made for Zopyrion, which have to be speculative at best,[31] there is

30 For an excellent summary of what these sources are and an introduction to the problems involved, see Dickey, (2007) 87–102.

31 The identity of this Zopyrion is a puzzling question: the name occurs in inscriptions from most parts of Greece (*LGPN* I-IIIB s.v.), but in literature the only promising references occur in Plutarch's

no way of verifying his contribution to the *On names and rare words*. The *Suda* (i.e., its source, Hesychius of Miletus), must have had his information from the same learned sources where all the lists of titles and other unverifiable biographical details come from. All we can say is that from a very early stage Zopyrion's contribution appears to have been neglected and the entire Lexicon credited to Pamphilus, because Athenaeus cites Pamphilus for words beginning with alpha on two occasions (ἄωτον, a type of cup, at *Deipn.* 11.783a, and ἀττανίτης, a type of cake, at *Deipn.* 14.645c). Moreover, if we accept the information about Zopyrion's involvement, we also have to accept an alphabetical arrangement for the entire work, something of which not all scholars are convinced.[32]

Returning to the fate of Pamphilus' lexicon after its initial publication, we note that the first abridgement was done by the 'sophist' Julius Vestinus:[33]

> Οὐηστῖνος Ἰούλιος χρηματίσας σοφιστής. Ἐπιτομὴν τῶν Παμφίλου Γλωσσῶν βιβλία ϟδ', Ἐκλογὴν ὀνομάτων ἐκ τῶν Δημοσθένους βιβλίων, Ἐκλογὴν ἐκ τῶν Θουκυδίδου, Ἰσαίου, Ἰσοκράτους καὶ Θρασυμάχου τοῦ ῥήτορος καὶ τῶν ἄλλων ῥητόρων.
>
> Vestinus, surnamed Julius; sophist. He produced an Epitome of Pamphilus' *Glossai* (94[34] books); *Selection of Words* from the works of Demosthenes; *Selection* from those of Thucydides, of Isaeus, of Isocrates and the orator Thrasymachus and the other orators. (*Suda* O 835)

Vestinus' sophistic credentials are supported by the rest of his lexicographical output, which was firmly focused on Attic authors and rhetorical speeches. Further information on his career can be found in an inscription that was originally in the Via Ostiensis in Rome but now survives only in a ninth-century manuscript in Switzerland (*Einsiedelnensis* 326).

> Ἀρχιερεῖ Ἀλεξανδρείας καὶ Αἰγύπτου πάσης Λευκί(ω)ι Ἰουλί(ω)ι Οὐηστίνωι καὶ ἐπιστάτηι τοῦ Μουσείου καὶ ἐπὶ τῶν ἐν Ῥώμηι βιβλιοθηκῶν Ῥωμαϊκῶν τε καὶ Ἑλληνικῶν καὶ ἐπὶ τῆς παιδείας Ἀδριανοῦ, ἐπιστολεῖ τοῦ αὐτοῦ αὐτοκράτορος.
>
> For the high priest of Alexandria and all Egypt Lucius Iulius Vestinus, head of the Museum and in charge of the Roman and Greek libraries in Rome and in charge of education under the em-

---

Sympotic Questions, where an elementary teacher (γραμματιστής) named Zopyrion expresses his opinions on the origin of the letters of the alphabet (738F–739 A), and is at a loss when asked about the wound inflicted by Diomedes on Aphrodite (739B-C). Such a man could conceivably have begun the production of an ambitious dictionary, and the *terminus ante quem* provided by Plutarch does make him a predecessor of Pamphilus'. Nothing certain can be said on his identity, however, particularly since both he and his interlocutors are otherwise unknown and do not figure elsewhere in Plutarch.

32 See most recently Tosi (2015) 625.

33 For more on his career as a sophist, see Bowie (2013).

34 All scholars adopt the number 95, given by the *Suda* at Π 142, as the correct number of books for Pamphilus' lexicon. Unlike the corruption from 95 to 405 books (see below), it is not clear how this corruption to 94 occurred.

peror Hadrian and the secretary of the same emperor. (CIG 5900 = *IG* XIV.1085 = *OGIS* 659, transl. Bowie)

Such a man, therefore, who had been in charge of the Mouseion in Alexandria and of the Libraries in Rome, would have had easier access than anyone to works inconvenient for reproduction and wide circulation, such as Pamphilus' 95-book Lexicon. It is not surprising that a man of such resources and with an evident interest in the Greek language became involved with what was one of the richest word-collections at the time, and sought to make it more accessible by means of an epitome. It is worth wondering whether Vestinus might have transported a copy of Pamphilus' lexicon to Rome, and thus made available this remarkable repository of Greek erudition in Roman learned circles. It is very hard to be certain of this, because there are no direct parallels specifically involving voluminous products of Greek scholarship.[35] At the very least Vestinus must have made a copy of his own epitome available. In the end, however, nothing survived from his efforts but a few scattered references. One of the reasons for this silence must be the success of another Epitome, this time by Diogenianus.

Διογενειανός, Ἡρακλείας ἑτέρας, οὐ τῆς Πόντου, γραμματικός, γεγονὼς καὶ αὐτὸς ἐπὶ Ἀδριανοῦ βασιλέως. [...] ἔστι δὲ αὐτοῦ βιβλία ταῦτα: Λέξις παντοδαπὴ κατὰ στοιχεῖον ἐν βιβλίοις ε′: ἐπιτομὴ δέ ἐστι τῶν Παμφίλου λέξεων βιβλίων ε′ καὶ τετρακοσίων καὶ τῶν Ζωπυρίωνος ...

Diogenianus, of the other Heraclea, not in Pontus.[36] Grammarian. He too lived under the emperor Hadrian. [...] His books are as follows: *Miscellaneous Lexis*, alphabetically arranged, in five books—this is an epitome of Pamphilus' lexicon in 405[37] books and of Zopyrion ... (*Suda* Δ 1140)

This epitome cannot have been composed much later than Vestinus', because Diogenianus is also dated within the second century CE. In fact, the *Suda* also places him in the reign of Hadrian, which makes him and Vestinus possible contemporaries.

This *Suda* article is the one and only instance where Diogenianus is mentioned in connection with Pamphilus, but there is additional evidence connecting Diogenianus and Vestinus: at the end of a Scholion in one manuscript (Laurentianus 8.7) of

---

35 There is, however, plenty of evidence of Greek books being brought to Rome or Italy, from Cicero and Atticus (see Cicero, *Ad Att.* 1.7; 1.10.4; 1.20.7), to Philodemus bringing an entire Epicurean library to Herculaneum (see Gigante [1995]), to Pliny the Younger (*Ep.* 1.8; 2.17). We also know that Galen possessed two Homeric texts as edited/corrected by Aristarchus, so they must have originated at Alexandria (*De indolentia* 13), and that Domitian sent scribes to Alexandria to make fresh copies or corrections for books that were destroyed by fires at Rome (Suetonius, *Dom.* 20.1).

36 The *Suda* mentions two more grammarians of the same name (Δ 1139, 1146). There is also a collection of proverbs attributed to Diogenianus, which survives in various recensions in a number of manuscripts, but its precise relationship to the lexicographical work is not clear. Hesychius did find proverbs listed (without context) in Diogenianus (cf. *Ad Eul.* 1.24).

37 This number must be due to a scribal error that confused the koppa, the symbol for ninety (ϟ), with the upsilon (Υ), which represents four hundred.

the sermons of Gregory of Nazianzus detailing the subdivisions of the *talanton* into *mnai*, *drachmai*, etc., we read:

> ὡς ἐν τοῖς Περὶ σταθμῶν Ἀπολλόδωρος, ἐκ τῶν Διογενιανοῦ τῆς ἐπιτομῆς τῶν Οὐηστίνου ἑλληνικῶν ⟨ὀνομάτων⟩...
>
> As Apollodorus says in his *On weight measurements*, from Diogenianus' Epitome of Vestinus' *Greek Words*... (Σ 71 Piccolomini,[38] Serm. 18.6).[39]

If this information may be trusted, then Diogenianus' *Miscellaneous Lexis* was a 'second-degree' epitome, using Vestinus instead of the original work by Pamphilus. Testifying against Diogenianus' epitome being based on Vestinus', is the statement of the *Suda*[40] and the short time span between them, because both epitomators are dated within the reign of Hadrian. Either way, these developments provide a very good case study for the way in which a voluminous work like Pamphilus' may attain some form of dissemination to a wider public, namely through (successive) abridgement.

Diogenianus' lexicographical output did in fact enjoy a long survival, and was used by later lexicographers and compilers of scholia down to Photius (ninth century) and the *Etymologicum Magnum* (until the twelfth century).[41] In all the cases where a gloss from Diogenianus is reported, the reference is very brief, sometimes with the addition of just one word (Διογενιανός, Διογενιανοῦ). The only testimonies that offer some more detail on the sources and rationale of his work are the ones discussed here. The following anonymous scholion on the *On types of style* by the rhetorician Hermogenes is especially important because it contains a very rare mention of a title for Diogenianus' work and serves as our only confirmation for the information of the *Suda* (Δ 1140), where the title given for the epitome of Pamphilus' Lexicon is Λέξις Παντοδαπή (*Miscellaneous Lexis or Vocabulary*):

> Φρύνιχος ἐν τῇ Σοφιστικῇ Παρασκευῇ παρατίθεται τὸ 'ὑπόξυλος ῥήτωρ' καὶ 'ποιητής' [...] μέμνηται καὶ Διογενιανὸς ἐν τῇ Λέξει τούτου καὶ Διονύσιος ἐν τῇ Ἐκλογῇ τῶν ὀνομάτων, ὅτι Μένανδρος ἐν τῇ Περινθίᾳ φησίν 'οὐδ' αὐτός εἰμι σὺν θεοῖς ὑπόξυλος'...

---

38 Piccolomini (1879) 244. These Scholia (or a nucleus therein) were compiled in Alexandria, possibly around the end of the 6th century, therefore in the same scholarly milieu as the Lexicon of Hesychius. The compiler(s) must have had access to Diogenianus, because the latter is also cited in another set of Scholia on the same author, and there are a number of parallels with Hesychius, who explicitly names Diogenianus as his principal source.

39 The same information is repeated in a late Scholion on *Iliad* 5.576, by a 12th-13th century hand on the 'Venetus B' manuscript (Bibliotheca Marciana gr. 821), where the reference to Vestinus is omitted: ἐκ τῶν Διογενιανοῦ τῆς ἐπιτομῆς ἑλληνικῶν ὀνομάτων ('from Diogenianus' Epitome of the *Greek Words*').

40 The *Suda* itself made use of the Scholia on Gregory (we even find the same information on *talanton* etc. at T 34), but the extent of this use cannot be ascertained.

41 Thus also Latte (1953) xliv.

> Phrynichus in his *Praeparatio Sophistica* cites the expression 'counterfeit orator' and '(counterfeit) poet' [...] Diogenianus also mentions this in his *Lexis* and Dionysius in his *Selection of Words*, says that Menander in the *Perinthia* writes: 'by god, I am not myself fake' (Anon. in Hermogenes Περὶ Ἰδεῶν, p. 486 Walz[42])

The scholion must be early or drawing on a generally well-informed source, because there are references to the Atticists Phrynichus and Aelius Dionysius, for whom titles are also given. The Patriarch Photius (ninth century CE) refers to 'the collection composed by Diogenianus' without giving a title and only from the contrast he draws with Helladius are we meant to understand that Diogenianus' work was focused primarily on poetic words, obeyed strict alphabetical order well beyond the first letters, and comprised five volumes of equal length, which is in agreement with the *Suda*'s report of a five-volume epitome of Pamphilus (*Bibliotheca* 145, 98b).

On the other hand, the Lexicon of Hesychius (composed in the fifth or sixth century CE in Alexandria), which is our principal witness on Diogenianus, seems to have been based on a different work from the one that was known to the sources of the *Suda* and the Hermogenes scholion. Its title was Περιεργοπένητες (*Periergopenetes*, '*Poor scholars*', or '*the Inquisitive poor*'), a strange title not found anywhere else, and certainly not in the reports that connect Diogenianus with Pamphilus (or Vestinus). In his dedicatory letter to Eulogius[43] which serves as the preface to his Lexicon, Hesychius tries to explain Diogenianus' title as significant of the author's intentions, namely to give poor people access to knowledge and help them satisfy their inquisitiveness and curiosity (*Ad Eulogium* ll.16–20).[44] We can reasonably connect this intention with the conciseness of Diogenianus' work (especially if it was an epitome of an epitome), and the 'stripping off' of much scholarly apparatus, quotations etc. This would gesture towards greater accessibility as an educational aspiration, although it is not clear how many truly poor people in the second century CE could afford (or would be genuinely interested in acquiring) their own copy of a lexicon of rare poetic and dialectal words. Moreover, Hesychius seems to think that he is dealing with an original work, in the sense of a compilation made from multiple earlier glossaries by Diogenianus himself, rather than with an epitome from a single major prototype.

> Πολλοὶ μὲν καὶ ἄλλοι τῶν παλαιῶν τὰς κατὰ στοιχεῖον συντεθείκασι λέξεις, ὦ πάντων ἐμοὶ προσφιλέστατε Εὐλόγιε· ἀλλ' οἱ μὲν τὰς Ὁμηρικὰς μόνας ὡς Ἀππίων καὶ Ἀπολλώνιος ὁ τοῦ Ἀρχιβίου· οἱ δὲ τὰς κωμικὰς ἰδίᾳ καὶ τὰς τραγικὰς ὡς Θέων καὶ Δίδυμος καὶ ἕτεροι τοιοῦτοι· ὁμοῦ δὲ πάσας τούτων οὐδὲ εἷς. Διογενιανὸς δέ τις μετὰ τούτους γεγονὼς ἀνὴρ σπουδαῖος καὶ φιλόκαλος, τά τε προειρημένα βιβλία καὶ πάσας τὰς σποράδην παρὰ πᾶσι κειμένας λέξεις

42 Walz (1833).

43 Identification is not secure, but an Eulogius Scholasticus active in the fifth century has been proposed.

44 Bossi (2000) interprets the title along the lines of an 'active' compound adjective, meaning something like 'caring *for* the poor'. He points to a late development in the meaning of περιεργάζομαι, 'to be busy about'.

συναγαγών, ὁμοῦ πάσας καθ' ἕκαστον στοιχεῖον συντέθεικε· [...] ἐπιγράψας τὰ βιβλία Περιεργοπένητας, καὶ ταύτῃ χρησάμενος τῇ διανοίᾳ· ἡγεῖτο γάρ, οἶμαι, μὴ μόνοις πλουσίοις, ἀλλὰ καὶ τοῖς πένησι τῶν ἀνθρώπων χρησιμεύσειν τε καὶ ἀντὶ διδασκάλων ἀρκέσειν αὐτά.

Many other old authors composed alphabetical dictionaries, my dearest Eulogius; but some collected Homeric words only, like Apion and Apollonius the son of Archibius; others collected separately words from comedy and tragedy, like Theon and Didymus and other such people; but not one person (had collected) all these together. A certain Diogenianus, however, a man who lived later than these authors and was sophisticated and a lover of beautiful things, brought together both the aforementioned books and the words scattered across all (authors) and composed a work listing them all together alphabetically. [...] He entitled his work *Poor scholars*, following this rationale. For he believed, I think, that this material would be useful not only to rich but also to poor people and would suffice instead of teachers. (Hesychius, *Ad Eulogium*, 1–19)

What is more, Pamphilus does not even figure among the sources of Diogenianus' *Periergopenetes* as described by Hesychius, which is striking, because distilled material from Pamphilus certainly *is* present in Hesychius. For instance, we can point to very close parallels between Athenaeus' extracts from Pamphilus that are cited above and the relevant entries in Hesychius: in the case of *epimelis* (*Deipn.* 14.650e) we read in Hesychius that it means 'brooch; and a type of pear' (E 4977); the liquid measurement *hemina* is glossed with the similar term *kotyle* (H 503) just as in Athenaeus' report on Pamphilus; the same applies to the entries cited above on *teuthis* (a cake, *Deipn.* 7.326e) and *hypoglottis* (a type of garland, *Deipn.* 15.677b) (cf. Hesychius T 685, Y 585).[45] These parallels indicate that even if the *Periergopenetes* was distinct from the epitome of Pamphilus/Vestinus entitled *Miscellaneous Lexis*, which is likely,[46] still Pamphilus was a major source for the *Periergopenêtes* too. This has two important implications: firstly, it means that Diogenianus included significantly overlapping material in his two works; and secondly it confirms that, via the lexicon of Hesychius and the use made of Diogenianus by later Byzantine lexicographers, Pamphilus left a lasting if indirect legacy in Greek lexicography. Why then is he not acknowledged by Hesychius? It could be that the debt was deliberately suppressed by Diogenianus himself, or it could be that no direct information about Diogenianus' sources was available to Hesychius, leaving him to postulate the use of Didymus, Apollonius, etc. from parallel texts that were more readily available to him than Pamphilus, who had fallen into a general oblivion after Athenaeus.[47]

45 The parallels are even more striking in those cases where the relevant words do not appear anywhere else in Greek literature. On *mathalides* (plural form for a type of drinking cup) compare μαθαλίδας ... Πάμφιλος δέ φησι "μήποτε ἐκπώματός ἐστιν εἶδος, ἢ μέτρον οἷον κύαθος" (Athenaeus, *Deipn.* 11.487c) and μαθαλίδες. ἐκπώματά τινα. οἱ δὲ μέτρα, ὡς κύαθοι. (Hsch. M 58). On *kodonea* or *kodonaia* (winter figs) compare Pamphilus at Athenaeus, *Deipn.* 3.77a with Hsch. K 3211. On *hydnophyllon* (a grass growing near/on top of truffles) compare Pamphilus at *Deipn.* 2.62d with Y 68.
46 Thus, e. g., Matthaios (2010) 175; for the opposite view, see Schironi (2009) 47.
47 Cf. Alpers (1981) 125.

Some more information on the modes of publication and dissemination of Diogenianus' work can be obtained from papyrological evidence.[48] As we saw above, many lexica on papyrus are elementary word lists associated more with school contexts than with Alexandrian erudition, but there are some specimens that can be clearly distinguished as scholarly products.[49] Two papyri from the fourth century CE, one from Oxyrhynchus and one of uncertain provenance (*P.Oxy.* 3329 and *PSI* VIII 892), have been associated with Diogenianus' *Periergopenetes*, whether they are copies of the work itself or represent different stages of excerption and re-compilation due to their very close affinities with the corresponding glosses in Hesychius. This connection has been laid out in detail by Esposito.[50] The papyri confirm the obvious point that Diogenianus' epitome had a much greater chance of circulation and survival than Pamphilus' 95-book magnum opus, with which no papyri can be associated with any confidence. However, as we do not know the social background of the original owners of *P.Oxy.* 3329 and *PSI* VIII 892, we cannot ascertain whether Diogenianus' stated publication aim of reaching a wider audience than the rich elite was realised in practice.

In any discussion of Greek lexica on papyrus, special mention must be made of *P.Oxy.* XV 1802 + LXXI 4812, another papyrus form Oxyrhynchus dated in the second or third century CE. In her recent detailed study of this text, Schironi concluded by also assigning this text to the tradition of Pamphilus-Vestinus-Diogenianus, mainly due to its strict alphabetisation (a feature of Diogenianus' work according to Hesychius and Photius, cf. above), the use of erudite sources also found in Athenaeus (such as Dinon's *Persica* and Aristotle's collection of *Constitutions*), and some exclusive parallels with Hesychius.[51] As Schironi points out, however, it is not possible to identify the papyrus lexicon with a specific stage of publication within the successive epitomisation process: the scholarly citations and lack of literary words speak against Diogenianus as the author. Additionally, the strong presence of Near Eastern languages is not consistent with Vestinus' *Greek Words*,[52] while the length and range of the entries is still too small to allow identification with Pamphilus' colossal work.

The following excerpts give an indication of the impressive scholarly citations in this papyrus lexicon, as well as its unique interest in non-Greek words from Near Eastern cultures (which included Lydian, Scythian, Persian, and Chaldaean):

μενεμανι τὸ ὕδωρ παρὰ τοῖς Πέρσαις. Δείνων (ζεινων pap.) ἐ[ν ¯ Περσι]κῶν.

48 Much progress has been made in the study of papyrus fragments of Greek lexica, whereas until fairly recently the main resource was Naoumides (1969), 181–202. Here I have relied mainly on Esposito (2009); Schironi (2009) and Ucciardello (2012).

49 Esposito (2009) 266.

50 Esposito (2009) 269–70.

51 Schironi (2009) 43–46.

52 Schironi (2009) 46–52.

menemani: water among the Persians. D(e)inon in Book ... of the *Persica*. (Fr. 3, ii 17, ed. and transl. Schironi)

μῆτραι ἐν Ταρσῷ καὶ Σόλοις τὰς δέλτους ἐν αἷς ἀπ[ογράφονται τὰς | οἰκίας μήτρας προσαγορεύ-εσθαι, ἃς καὶ δημ[οσίας. Ἀριστοτέ|λης ἐν τῇ Σολέων πολιτείᾳ.

metrai: in Tarsus and Soloi writing tablets on which they register houses are called 'metrai', and they (are) also public. Aristotle in the *Constitution of Soloi*. (Fr. 3, iii 5 – 7, ed. and transl. Schironi)

Given the remarkable content of this lexicon, one hypothesis might be that it was the result of a scholar (or network of scholars) making their own 'personalised' use of Pamphilus' lexicon (and perhaps other sources as well), by selecting and excerpting what they considered to be of particular importance to their own work. In the case of *P.Oxy.* 1802 + 4812 a principal point of interest was information about Near Eastern cultures and their languages, which we could speculatively associate with the presence of a history of Alexander on the recto of the roll (the lexicon was written on the back).

The presence of such specialised scholarly material in the rubbish heaps of Oxyrhynchus leads to important questions about their users, and the ways in which they accessed and interacted with the scholastic culture that was centred in the library of Alexandria. We saw above some early indications for how individual users might have disseminated the products of Hellenistic scholarship outside the capital. We have much more information about the Roman period, largely because of the chronological range of the majority of papyrological finds. We even have texts referring explicitly to the production and exchange of books. As far as the circulation of *scholarly* material in Roman Egypt is concerned, a letter found in Oxyrhynchus, *P.Oxy.* XVIII 2192, dating from the second half of the second century CE, is particularly revealing. We should recall that Oxyrhynchus was an important provincial city in the second century CE and had close connections with Alexandria. Many prominent Alexandrian citizens owned property at Oxyrhynchus and held municipal offices there.[53] It is therefore reasonable to expect that valuable items, including books, would be moving between the two places.

The letter in question provides some insight precisely into this movement. Its main interest lies in the two postscripts (in two different hands), where what must be two separate individuals with advanced scholarly interests discuss the acquisition of specialist items such as books 6 and 7 of the *Komodoumenoi* (*Comedy characters*) by Hypsicrates, the *Epitomes of tragic myths* by Thersagoras (col. ii. 28 – 30, 35 – 36), and a multi-volume grammatical work, perhaps entitled *Tenses*, by Seleucus (col. ii. 42). Given that the letter was found in Oxyrhynchus, it is somewhat puzzling that it mentions the prominent Alexandrian lexicographers Harpocration and Valerius Pollio as located close to the recipient rather than the sender (for example,

53 Turner (1952) 85.

the recipient is advised to look for the *Comedy characters* among Pollio's books, on the advice of Harpocration, col. ii. 30–33).[54] Whatever the precise details of the respective locations and relationship between sender(s) and recipient, the letter is extremely informative about the ways in which these specialist scholarly works circulated in second-century Egypt, and the identification of the individuals involved 'puts Oxyrhynchus in contact with the best Alexandrian scholarship of the second century'.[55] The text encourages us to think of networks connected by personal bonds of friendship, where scholars would procure copies from one another (note the expression ποιήσας μοι πέμψον, 'make (copies) and send them to me', at col. ii. 29–30 and 43–44), hence it was important to know which members of the network possessed the relevant books. In our letter the private collections identified are those of Valerius Pollio and his son Diodorus (col. ii. 32–33). At the same time, however, we are told that some of the material may (also) be found at the shop of a bookseller by the name of Demetrius (col. ii. 37), which means that the book trade also played a role in the process of publication and dissemination of these products of ancient erudition in Roman Egypt.[56]

Conclusions

Although it is not always easy to align the indirect tradition of the fragments of Hellenistic lexicography that are preserved by later authors with the direct evidence of papyri, we have seen evidence of at least a 'trickle down' effect whereby work done in the privileged surroundings of the library was able to influence reading and educational practices in the wider community. This is not to deny that, just like today, the main audience for scholarly works were like-minded scholars, often all working in the same library. In such circumstances, publication would consist simply in depositing a copy among the holdings of the library. On the other hand, we have seen evidence that the work of Hellenistic lexicographers was also made available through some alternative form of publication that brought it beyond the 'ivory tower'. From the early Hellenistic period onwards, the final outcomes of this process were elementary word lists that served reading and teaching purposes. The erudite material and parallels with Hesychius noted by Uciardello in *P.Berol.* inv. 9965 (see above) allow us to postulate an impressive degree of continuity across many centuries of Greek lexicography.

---

54 The identifications of these individuals are central to the Alexandrian connection, which is denied by Johnson (2010) 180–83. For more on these issues see Turner (1952); Hatzilambrou (2007) 282–86. I have also relied on Hatzilambrou for the text of the letter.

55 Turner (1952) 92.

56 See Johnson (2010) 183–85 for further discussion of papyri indicating comparable exchanges about specialised book resources.

Pamphilus' work must have represented a pivotal phase of this development. Its subsequent fate is emblematic of the price paid for wider dissemination, namely dramatic reductions in size, excerption and selection,[57] as well as frequent combination and re-compilation with other sources. We have seen, for example, that the compiler of *P.Oxy.* XV 1802 + LXXI 4812 was especially interested in non-Greek words, while Diogenianus (according to Photius) focused on poetic words. Since the responsibility for post-publication reproduction often lay with the reader, the form and content of circulating books could be dictated by users' interests and priorities. Networks of scholars with members residing both at Alexandria and in the provinces and specialist booksellers must have played a key role in this process, as we saw in the exchange in the postscripts of *P.Oxy.* XVIII 2192. This form of publication had the surprising (by modern standards) consequence that concern for preservation of the original wording was not a priority, especially since the works of Greek scholarship were 'open' texts,[58] where extractable information was valued infinitely more than the texts' own style or potential literary merit.

57 This phenomenon was by no means restricted to the Greek scholarly world. See the contribution by Catherine Hezser in this volume.
58 This idea is developed in Sluiter (2000b).

Matthew Nicholls

'Bookish Places' in Imperial Rome: Bookshops and the Urban Landscape of Learning

The second-century medical author Galen, at the beginning of his treatise 'On My Own Books', gives a lively picture of book-based literary dispute in central Rome. The setting is not one of the great libraries or lecture halls nearby – though Galen certainly also used and talked about these – but among the commercial booksellers of a nearby street:

> ἐν γάρ τοι τῷ Σανδαλρίῳ, καθ' ὃ δὴ πλεῖστα τῶν ἐν Ῥώμη βιβλιοπωλείων ἐστὶν, ἐθεασάμεθά τινας ἀμφιβητοῦντας εἴτ'ἐμὸν εἴη τὸ πιπρασκόμενον αὐτὸ βιβλὶον εἴτ'ἄλλου τινός· ἐπεγέγραπτο δὴ γὰρ, Γαληνὸς ἰατρός· ὠνουμένου δέ τινος ὡς ἐμὸν ὑπὸ τοῦ ξένου τῆς ἐπιγραφῆς κινηθείς τις ἀνὴρ τῶν φιλολόγων ἐβουλήθη γνῶναι τὴν ἐπαγγελίαν αὐτοῦ· καὶ δύο τοὺς πρώτους στίχους ἀναγνοὺς εὐθέως ἀπέρριψε τὸ γράμμα, τοῦτο μόνον ἐπιφθεγξάμενος ὡς οὐχ ἔστι λέξις αὔτη Γαληνοῦ καὶ ψευδῶς ἐπιγέγραπται τουτὶ τὸ βιβλόν.
>
> I was recently in the Vicus Sandaliarius, the area of Rome with the largest concentration of booksellers, where I witnessed a dispute as to whether a certain book for sale was by me or someone else. The book bore the title *Galen the Doctor.* Someone had bought the book under the impression that it was one of mine; someone else – a man of letters – struck by the odd form of the title, desired to know the book's subject. On reading the first two lines he immediately tore up the inscription, saying simply: 'This is not Galen's language – the title is false.' (Galen, *On My Own Books* 1)[1]

Since the disputed author was Galen himself, he was no doubt able to make a decisive personal intervention; his account of the incident leads to a discussion of intellectual formation and literary attribution, and opens a written work aiming to establish an accurate canon of his own circulating writings.[2]

We can also consider what this story tells us about the commercial and public character of this part of Rome as seen from the street, a topic of increasing interest

This chapter was given as a paper at the conference 'Ancient Scholarship: Scholastic Culture in the Hellenistic and Roman Eras', held at University of Glasgow in April 2017. I am very grateful to the organisers, Sean Adams and Zanne Domoney-Little, for their invitation to speak, and help and patience with this contribution, and to the other delegates (including authors in the present volume) for their useful and generous suggestions.

1 Galen, *Libr. Propr.* 1.1 (K.19.8); translation by Singer (1997) 3.
2 For a discussion of the intellectual background to this discussion and its implications for the culture of reading and education in Rome, see Johnson (2010) 85–97.

Matthew Nicholls, University of Reading

https://doi.org/10.1515/9783110660982-005

to Roman historians.[3] The street in question is the Vicus Sandaliarius, which ran to the north-east of the Templum Pacis.[4] This was an important street in the road system leading towards the Forum, imperial fora, and Palatine from the eastern side of the city; we will return to the significance of those monumental areas later.

The street itself, however, had a commercial character. Originally named for shoemakers,[5] it was by Galen's time, as we have just seen, the place where 'most of the booksellers in Rome are'. At its northern end near the Porticus Absidata, a monumentalising point of articulation between the Forum Transitorium and roads leading from the densely populated residential districts of the Suburra and Esquiline, it joined another street, the Argiletum, which was also known for its booksellers.[6] At its other end, to the southeast, it pointed towards the Baths of Trajan and the possible location of the Sigillaria, named twice by Aulus Gellius as an area where antiquarian booksellers had their shops.[7] On the other side of the Forum area, to the south, the Vicus Tuscus is also implied by Horace as a place where booksellers could be found.[8] The commercial streetscape in the immediate vicinity of the Forum therefore seems to have been known for its provision of bookshops, at least to bookish authors, writing on literary and intellectual matters for an interested audience; all the examples that we know about come from this relatively small area, which we will explore as an urban landscape containing various sorts of 'bookish place'.

The customers stopping to browse at these bookshops became part of the neighbourhood's intellectual character, as expressed in its street life. This is apparent in Galen's anecdote, in which he was able to notice an informal and animated literary discussion, whether inside the shop itself or in an area in or visible from the street. The participants were not known to Galen, or (it would seem) to each other; it was the quasi-public commercial public space of the bookshop that had brought them together and created the context for an intellectual encounter, a range of impromptu oral responses to a written text. Galen's anecdote serves a particular purpose within his treatise, leading into an account of his own works aimed at avoiding or resolving pseudepigraphical claims, but there is no reason to doubt the verisimilitude of this opening story. The urban context in which he sets it seems credible; the story needs a plausible setting to carry weight, and as it happens Galen's near-contemporary Aulus Gellius says something very similar about exactly the same street:

3 See for example Hartnett (2017); Östenberg *et al.* (2015); Laurence/Newsome (2011); Holleran (2012).

4 Richardson (1992) s.v. Vicus Sandaliarius (p. 427); cf. Coarelli/Steinby (1999) s.v. Vicus Sandaliarius (p. 189), and Carandini/Carafa (2017): Vol. 2 table at 20 (where it is spelled Vicus Sandalarius).

5 Martial, *Ep.* 2.17 implies that the neighbouring Argiletum was also still used by shoemakers in the late first century CE. On the book trade in Rome, see Winsbury (2009) ch. 6; White (2009).

6 Argiletum: Martial, *Ep.* 1. 2, 1. 3, 1.117. Cf. Catullus 55.4 for implied clustering of book shops.

7 Sigillaria bookshops: Gellius, *NA* 2.3.5, 5.4.1. Possible location in Trajan's Baths: *Schol. ad Iuv.* 6.154.

8 Horace, *Ep.* 1.20.1–2.

in Sandaliario forte apud librarios fuimus, cum ibi in multorum hominum coetu Apollinaris Sulpicius, vir in memoria nostra praeter alios doctus, iactatorem quempiam et venditatorem Sallustianae lectionis inrisit inlusitque...

I happened to be in the (Vicus) Sandaliarius among the booksellers just when, in the throng of many people there, Apollinaris Sulpicius, a man more learned than any other I have known, was making fun of some boastful fellow parading his reading of Sallust... (Gellius, *NA* 18.4.1)

Here the impression of the gathering from Galen's anecdote is amplified by reference to a 'throng of many people': this sounds like a busy place. The two stories share common themes. They both concern the capacity for learned individuals (the author, and/or his circle of learned friends) to provide a definitive answer to questions of textual authenticity, suggesting that the authenticity of the contents of bookshops (and the judgement of some of their customers) were sometimes open to dispute. But we should also note that both stories feature gatherings which included people not already known to each other and the presence of genuine experts – Galen, Gellius, Sulpicius Apollinaris – who either passed by these bookshops or went there deliberately as part of their practice as public intellectuals.

In this mix of people and discussion, both stories suggest that the bookshops of the Vicus Sandaliarius provided a space for the sorts of conversations which were also located by these same authors inside libraries – discussions about textual authenticity, attribution, points of grammar, and so on, and that libraries and bookshops thus shared some perceived characteristics as 'bookish places' in the urban landscape.

In the following passage, for example, the very same cast of characters – Gellius and Sulpicius Apollinaris – enact a similar demolition job on the opinions of another reader as their group sits in the library of the Domus Tiberiana:[9]

cum in domus Tiberianae bibliotheca sederemus ego et Apollinaris Sulpicius et quidam alii mihi aut illi familiares, prolatus forte liber est ita inscriptus: M. Catonis Nepotis. tum quaeri coeptum est, quisnam is fuisset M. Cato Nepos. atque ibi adulescens quispiam, quod ex eius sermonibus coniectare potui, non abhorrens a litteris: 'hic' inquit 'est M. Cato, non cognomento Nepos, sed M. Catonis Censorii ex filio nepos...'

When I was sitting in the library of the Domus Tiberiana with Sulpicius Apollinaris and some others who were friends of his or mine, a book happened to be brought to us with the label 'by Marcus Cato Nepos'. So we began to inquire who this Marcus Cato Nepos was. At that point some young fellow, not unacquainted with letters, so far as I could judge from his language, said: 'This is the Marcus Cato who was not called Nepos as a surname, but as the grandson of Marcus Cato Censorius through his son...' (Gellius, *NA* 13.20)

9 For Gellius' accounts of similar learned discussions in library spaces, see e.g., *NA* 5.21.9 (books in the Templum Pacis); 18.9 (a manuscript in a library at Patrae); 9.14 (a Quintus Claudius text in the library at Tibur); 19.5 (the library at Tibur settles a learned dispute).

Here an imperial library on the Palatine is the setting for the same kind of discussion, among the same people, as both Gellius and Galen set among the bookseller's shops on the nearby Vicus Sandaliarius. The aim of this chapter is to suggest that this similarity between the types of people and activities seen in bookshops and libraries explains the concentration of the former in an area close to the latter: that the cluster of booksellers was a market-driven phenomenon at least partly responding to an imperial euergetistic phenomenon, the dense provision of spaces for intellectual activity, including public libraries and lecture halls, in this same part of the city. The people attracted by these facilities were also likely customers for booksellers, leading the latter to cluster in an area of the city where they were most likely to thrive.

Scholars are increasingly interested in the commercial spaces of Roman streets.[10] While the nature of commercial spaces in Rome must have varied hugely, the descriptions in Gellius and Galen make it sound like we are dealing here with fixed premises holding a stock in a static location (the Vicus Sandaliarius), which could appeal to and accommodate casual passing trade—that is, something we might reasonably describe under the broad term 'shop'.

The evidence from places like Pompeii, the marble plan at Rome, and the iconography of Roman shopkeepers' tombs suggests that the most likely physical configuration for such shops would have been rows of single-room spaces opening onto the street through wide shuttered doorways. We can turn to literary descriptions to repopulate these cellular architectural spaces with human actors and displays of stock that blended indoor and outdoor, private and public space.

The goods on display in Roman shops seem often to have spilled over into the street, with the obvious aim of attracting passing custom. Martial famously describes the encroachment of commercial activities into the street before Domitian put a (doubtless temporary) stop to it:

> *abstulerat totam temerarius institor urbem*
> *inque suo nullum limine limen erat.*
> *iussisti tenuis, Germanice, crescere vicos,*
> *et modo quae fuerat semita, facta via est.*
> *nulla catenatis pila est praecincta lagonis*
> *nec praetor medio cogitur ire luto,*
> *stringitur in densa nec caeca novacula turba,*
> *occupat aut totas nigra popina vias.*
> *tonsor, copo, cocus, lanius sua limina servant.*
> *nunc Roma est, nuper magna taberna fuit.*

The audacious retailers had appropriated the entire city; no threshold kept within its own bounds. You bade the narrow streets expand, Germanicus, and what had lately been a track became a road. No column is girt with chained flagons, and the praetor is not forced to walk through the mud. The razor is not drawn blindly in a dense crowd, nor does the grimy cook

10 See note 3 above.

> shop monopolize the whole street. The barber, the taverner, the cook, the butcher keep to their own thresholds. Now it is Rome, it used to be a big shop. (Martial, *Ep.* 7.61)[11]

This sort of interaction with the space of the street seems to have happened in the bookshop stories told by Galen and Gellius, both of which describe encounters that took place in the street or were at least visible to the authors as they passed by; the risk to fragile and valuable books from such displays had to be balanced against the advantages of advertising to passers-by and showing the stock in good lighting conditions. The commercial displays of the bookshop keepers therefore appear to have been appealing speculatively to a passing trade through display and advertisement (rather than, say, catering exclusively to a private clientele by appointment, or by visiting them in their own homes, though this may also have been part of their business). The Gellius anecdote mentioned above is one of several in the *Noctes Atticae*; two of the others, at 5.4.1 and 13.31.1, mention characters 'sitting' in the shops, implying that there was also space and furniture for more lengthy perusal of the stock.

We can get some idea of what these bookshops might have looked like from Martial:

> ... *Argi nempe soles subire Letum:*
> *contra Caesaris est forum tabernas*
> *scriptis postibus hinc et inde totis,*
> *omnis ut cito perlegas poetas.*
> *illinc me pete. †nec† roges Atrectum—*
> *hoc nomen dominus gerit tabernae—*
> *de primo dabit alterove nido*
> *rasum pumice purpuraque cultum*
> *denaris tibi quinque Martialem.*
> *'tanti non es' ais? sapis, Luperce.*

> ... no doubt you often go down to [*the*] Argiletum. Opposite Caesar's Forum there's a shop with its doorposts completely covered by advertisements, so that you can read the entire list of poets at a glance. Look for me there. Ask for Atrectus (that being the name of the shop's proprietor), and he will hand you from the first or second pigeonhole a Martial, shaved with pumice and smart with purple, for five denarii. "You're not worth it," you say? You're a man of sense, Lupercus. (Martial, *Ep.* 1.117.9 – 18)[12]

Here is a physical, permanent shop premises 'opposite Caesar's Forum' in the Argiletum, the street intersecting with Galen's Vicus Sandaliarius that we met earlier. Its proprietor is named to ensure that the reader can find it in a city without numbered

---

11 Shackleton Bailey (1993).
12 Shackleton Bailey (1993). Naturally the extent to which we can rely literally on Martial's accounts, or their apparent character as self-advertisement, is open to dispute. The same argument is suggested as above for Gellius and Galen: some element of verisimilitude would strengthen the author's overall case.

street addresses. The shop has some sort of public display of the books on sale that covers its doorposts: that is, a display that deliberately addresses passing trade and attempts to attract custom by something that we might think of as advertising (we can compare references in Horace to the display of the latest poetry, and perhaps in Seneca; for a surviving archaeological example from a different kind of business, see the extraordinarily vivid street facade of the fullery of Ululitremulus in Pompeii, which competed for attention with a bright polychromatic chequerboard shop-front, frescoes of Romulus and Aeneas with parodic Virgilian poetry, and, perhaps, its memorable name).[13]

Martial's epigram also mentions numbered *nidi*, pigeon holes, containing the books. Ancient book shelving and collocation practice is poorly understood: there is little unambiguous literary testimony and the physical infrastructure of shelves and labels has all but disappeared.[14] Martial's reference to *nidi* is therefore precious, and can perhaps be taken to imply a system of niches similar to that found in the archive building at Dura Europos, mentioned in Gaëlle Coqueugniot's paper in this volume.[15] It would be a stretch to interpret Martial's description as a ranked display of Roman best-sellers, but it sounds like an orderly and predictable stock kept for speculative custom attracted from among the people passing in the street, rather than made to order for a known clientele.

A picture is emerging of public-facing establishments designed at least in part for passing trade and browsing – genuinely commercial entities, operating in a recognisable, economically rational way. In this they fit well into the wider understanding of Roman shops, shopping and streetscape in recent works on the subject.[16] Bookshops do not seem from the available evidence to have been exceptional in their layout, operation, or deliberate address to the customer: books, by the imperial period, were just one commodity among many bought and sold in Rome from shop premises facing onto the city's streets.

Behind the shop fronts we can assume facilities for production, whether to order or for stock. A well-known relief carving from Ostia gives us a suggestion of what such a Roman copying workshop may have looked like. It shows two writers seated at horizontal tables (a surprisingly rare attestation of what might seem an obviously

---

13 Horace, *Serm.* 1.4.71 and *Ars* 373. Cf. Seneca, *Ep.* 33.3 on deceptive *oclifera*, alluring street signage, but he is not explicitly talking about bookshops here. The Fullery of Ululitremulus is at insula IX.13.4–6 in Pompeii and its reconstructed façade, painted by Alberto Sanarica, is pictured in Hartnett (2017) Pl. VI.

14 Our best chance of observing shelved library books *in situ* vanished with the undisciplined excavation of the Villa of the Papyri in Herculaneum, where the books were apparently found on and among wooden shelf cabinets. See the contemporary account preserved under the names of Cerati/Locke (1755) 112–115; Sider (2005) 16–23.

15 See G. Coqueugniot's article in the present volume, especially Fig. 2.2 on p. 10, and Coqueugniot (2012) 97–98. I am grateful to Dr Coqueugniot for her generous correspondence on this fascinating site.

16 See e.g., Hartnett (2017) 60–67; Holleran (2012) 194–231.

convenient way of using ancient writing technology; ancient depictions more often show seated readers holding scrolls without a table).[17] The writers are apparently taking notes into bound wooden writing tablets while a centrally placed togate figure holding a scroll and standing on a podium makes a gesture of address. An audience sitting behind the writers exhibits a lively reaction. The scene seems to depict a lecture or address of some sort at which the spoken word is being converted into the written word, or possibly copyists taking dictation (an efficient way of producing more than one copy of a book at once).

There is little indication of sub-specialisation among Rome's booksellers. The work mentioned in Galen's story is a medical text (in Greek, given its putative attribution to Galen); Gellius' anecdote is about a disputed reading of the Latin historian Sallust; Martial tells us about the display and sale of Latin poetry books. It is possible that shops specialised in one subject area or another, but the sources do not say this and the generalist character of the bookshops echoes what we know about the nearby libraries, none of which have a discernible thematic specialism. Scribal labour (whether slave or free) would have been among the booksellers' greatest running costs, so a degree of specialism, whether by language or by subject area, might seem reasonable insofar as it would tend towards greater efficiency, given the economically rational behaviour we have already noted.

A linguistic staff specialisation is in fact echoed by the funerary epigraphy of Rome's library attendants, who sometimes mention an attachment to a Latin or a Greek library, but I have suggested elsewhere that this distinction (which largely disappears from the epigraphic record by the mid first century CE) need not be interpreted to imply physically separate library spaces for the two languages,[18] and the same is true of the bookshops here: we do not know of any specialist Latin or Greek booksellers in Rome.

Serena Ammirati's chapter in this volume gives fascinating details of the physical features of Latin legal texts in this period (such as the regular use of rubrics for legal texts only, and the distribution of text in *capita*, often highlighted by *ekthesis* and enlarged initial letters.)[19] as she notes, other types of text, such as Vitruvius' illustrated work on architecture or the artistic layout of poetry collections, would also have needed particular layouts and therefore particular scribal skills. The market seems to have developed far enough to sustain a population of booksellers who

---

17 Ostia inv. 130. See Del Corso (2009) at Fig. 17 and page 327. My thanks to Prof. A. Stramaglia for providing a copy of this article. On the rarity of horizontal tables, see e.g., Nicolai (2000), 'la presenza di tavoli per la consultazione, prevista da alcune ricostruzioni moderne, è giustamente messa in dubbio.' NB also the portable lecterns or scroll holders (and the occasional obliging Muse) depicted as alternative reader aids in Roman painting and relief art: see e.g., Knauer (1993) esp. 18–26 with figs. 25, 26, 30 ('reading and writing while seated at a desk or table do not seem to have been generally favoured before the Carolingian age', p. 26); cf. Martial, *Ep.*14.84 for a wooden book holder.
18 Nicholls (2011) 11–21.
19 See the concluding paragraphs of S. Ammirati's chapter in the present volume, p. 80.

made a living from general sales to a passing public; to argue that the economically specialist cluster of booksellers discussed here contained further sub-specialists adept in these various areas would be to go beyond our evidence, though the possibility remains.

So far, we have seen that imperial Rome contained several bookshops, whose location, stock and probable appearance were intended to attract passing customers; the presence and interaction of those customers is reported by various ancient authors. We have also seen that all the bookshops whose locations we know were in the centre of town, around the Forum area, and particularly in the Argiletum and latterly the Vicus Sandaliarius. It is to this clustering that we now turn.

There is a fair amount of evidence for retail clustering of this type at Rome.[20] Although it may at first glance seem perverse for businesses to try to locate themselves alongside competitors, the advantages are fairly obvious, especially for sellers of relatively high-value 'comparison goods' whose buyers like to shop around and think carefully about their purchase, like the customers in Galen and Gellius' stories. Research suggests that such customers spend more if they can make their comparisons easily,[21] and the frequent clustering of high-end shops like jewellers, tailors, and estate agents in modern cities would seem to bear this out: topographical clustering by specialisation allows buyers and sellers to find each other and encourages a competitive market for sales that ultimately makes the whole industry more sustainable.

On the production side, the co-location of similar businesses gives them similar competitive advantages in their own market for raw materials and supplies, helping to lower overheads, and allowing for pooling of expertise and market knowledge and the achievement of economies of scale in staffing, the development of specialist localised chains of production (where different workshops work on different elements of a product), and the lowering of transport costs. Some of our neighbourhood's other 'bookish spaces', its large public libraries, may well have had similar needs, resources, and staff for maintaining their own stocks[22] (they may also have constituted another market for the booksellers, though it seems unlikely that their own stocks were maintained by retail purchases). If so, the clustered provision both of specialised labour like the *librarioli* who acted as *glutinatores* (gluers, or binders) and bookshelf carpenters in the private book collections of Cicero and Atticus,[23] and of their specialised tools, raw materials, and market expertise, could have provided an econ-

20 Holleran (2012) Ch. 1 and esp. 51ff., 'The topography of distribution'.

21 Goodman (2016) 307–308.

22 Compare the endowment of the second-century CE library of Celsus at Ephesus. The interest on the sum of 23,000 denarii was intended to pay for staffing and book purchases, suggesting an envisaged investment in maintaining and expanding the book stock over time. Platthy (1968) 155–56 no.128.

23 Cicero, *Att.* 4.4a.1; 4.5.3.

omy of scale and contributed further to the clustering effect suggested here.[24] The majority of the criticism levelled at commercial booksellers by ancient authors concerns the poor quality of their wares (both the Galen and the Gellius anecdotes above turn on disputed authenticity, and neither invokes the acumen of the bookseller himself as an authority);[25] though that reputation might result from ignorance or actual dishonesty in passing off forgeries, we might also suspect that it resulted from the economic tension between lowering the cost of in-house production and maximising sale prices, and therefore interpret the clustering of booksellers as an economically rational desire to minimise the costs of supply, labour and travel as well as to maximise exposure to customers.

Street names in ancient Rome reflect this tendency for retail and service industries to cluster; Holleran notes that the fourth-century regionary catalogues preserve a Vicus Vitrarius (glassmakers), a Vicus Unguentarius (perfumers), and an Area Pannaria (second-hand cloth). Inscriptions give us the Vicus Lorarius (harnessmakers), the Vicus Frumentarius (grain dealers), and the Vicus Materiarius (carpenters or timber merchants).[26] This phenomenon is particularly clearly noted in ancient Rome for luxury or high-value trades, as might well be expected: every neighbourhood required a supplier of day-to-day goods, and high volume from a predictable market could make up for the comparatively low value of the goods sold.[27] Rare purchases of higher-order goods, required by a smaller proportion of the city's population and bought much less frequently, could be served well by a clustered topographical distribution which can be seen in the same parts of Rome as the booksellers. The Vicus Tuscus (where we have seen that Horace also located bookshops) was home to traders in silks, purple dyed cloth and perfumes.[28] The Via Sacra, another of the routes leading towards the Forum not far from the Vicus Sandaliarius, was particularly well known for its cluster of gold and jewellery workshops. A series of funerary inscriptions for its *aurifices* and *margaritarii* name the location of their businesses, well placed to attract the attention of wealthy clients making their way to and from the Forum and Palatine, as if the Via Sacra neighbourhood was itself part of their pro-

24 Compare a similar cluster in medieval Paris, mentioned by Peter White. Here booksellers profited from their 'close proximity to court, church and university circle who made up their clientele' and also from the fact that 'it was ... an area in which tradesmen involved in the production of individually commissioned books – parchment and paper dealers, scribes, illuminators, and binders – were already established'. White (2009) 272, citing Rouse/Rouse (2000) 1.11–49.

25 For example: Strabo, *Geogr.* 13.1.54, Cicero, *Q.Fr.* 3.4.5, 3.5.6, Quintilian, *Inst.* 9.4.39.

26 Holleran (2012) 54.

27 Though there is some evidence for clustering even of artisans involved in the production of that most quotidian of products, bread. Coates-Stevens (2006) has suggested that the Porta Maggiore area might have been home to a baking industry with storage dolia and flour mills; Laurence (2007) 55–57 has shown that bakeries tended to proliferate along major roads, suggesting that access to customers was important, though their relatively wide distribution through the town also suggests the importance of neighbourhood supply.

28 Holleran (2012) 246.

fessional identity.[29] There were also dealers in flowers and fruit there:[30] given the difficulty of maintaining a regular supply of seasonal highly perishable goods to the centre of town without the benefit of mechanised or refrigerated transport, and the probable high property costs of this prestigious location, this is good evidence for the 'pull' factor of a concentrated ready market for relatively expensive goods. The Saepta Iulia in the Campus Martius also became a centre for the luxury goods trade.[31] Martial gives us an amusing picture of a would-be connoisseur, Mamurra, who spends hours browsing there among the bronze statues, precious drinking cups, slaves, ivories, furniture and jewellery before eventually buying two cheap drinking cups for an *as*.

While some of the trade in luxury goods must have occurred in the homes of the wealthy, with merchants calling to offer their wares or take commissions, the presence of these clustered, specialist retail trades in these city centre areas indicates that the density of passing traffic, and the benefits of storing and displaying a certain volume of stock, made it profitable to establish retail premises serving the public, as we have seen for books. Moreover, Martial and others suggest that being seen in luxury shopping locations, displaying acumen and connoisseurship as a buyer, was a marker of taste, and this too seems to have been part of the point of the bookshop stories from Galen and Gellius with which we opened.

By clustering along the Vicus Sandaliarius and Argiletum, then, Rome's booksellers were behaving in a rational manner that can also be seen for other trades. Why were they in this particular location? Part of the answer must simply have been that they needed a good concentration of wealthy customers, like the jewellers and fruit sellers a short distance away in the Via Sacra.[32] But that is not the only answer; other parts of Rome, like the Saepta, also developed a luxury goods market, but books were particularly concentrated in this area. Moreover, not all the interlocutors in the Galen and Gellius stories are known to each other, well to do, or well connected, as one might have expected if regular access to a small, elite population was all that mattered.

I suggest that the answer lies in the character of some of the spaces and buildings in the immediately adjacent monumentalised heart of Rome, in the area around the Forum, imperial fora, and Palatine. In the remainder of this chapter I will consider briefly five different but mutually interlinked types of intellectual space that joined the bookshops discussed above in contributing to the character of this part of Rome. Nearby official institutional buildings dedicated by their imperial patrons

29 Palombi (1990). E.g., *CIL* 6.9207 = ILS 7685: M(arcus) Caedicius Iucundus | aurifex de | sacra via vix(it) a(nnos) XXX || Clodia ... Cf. *CIL* VI 9221 = *ILS* 7694, 9239, 9418 = *ILS* 7700, 9419, 9545 = *ILS* 7602, 9546–49.

30 Flowers: *CIL* VI.9283; fruit: Ovid, *Ars Am.* 2.265–66.

31 *ubi Roma suas aurea vexat opes:* Martial, *Ep.* 9.59.2.

32 This point was the subject of an interesting discussion following the presentation of this paper in Glasgow, and I am grateful to Prof. Catherine Steel for her useful observations.

to literary, documentary and intellectual activity included libraries, archives, and auditoria. These drew in various sorts of intellectual visitor and thus altered the commercial landscape of the surrounding streets,[33] for booksellers (as we've seen) and also (as below) for a fourth type of facility or activity, personal storage accommodation rented on a commercial basis, and a fifth, education. Each merits a fuller discussion that space permits here; my intention is simply to sketch out their place in the topography of this part of the city as factors fed by and contributing to the presence of the sorts of people seen in Galen and Gellius' stories, and the consequent clustering of book sellers.

Galen, whose encounter among the Vicus Sandaliarius bookshops was our starting point, was also a prolific user of Rome's imperial public libraries, and gives us the clearest demonstration of the incorporation of private and public space and resources in his account of his intellectual practice in this part of town. The recently rediscovered text of his *Peri Alupias* has given us remarkable insights into the operation, contents and use of these libraries before their destruction in the fire of 192 CE:[34]

> Διεφθάρη δὲ νῦν τὰ μὲν ἐν τῷ Παλατίῳ κατὰ τὴν αὐτὴν ἡμέραν τοῖς ἡμετέροις, τῆς πυρκαϊᾶς οὐ μόνον ταῖς κατὰ τὴν ἱερὰν ὁδὸν ἀποθήκαις λυμηναμένης ἀλλὰ καὶ πρὸ αὐτῶν μὲν ταῖς κατὰ τὸ τῆς Εἰρήνης τέμενος, μετὰ ταῦτα δὲ ταῖς κατὰ τὸ Παλάτιον τε καὶ τὴν Τιβεριανὴν καλουμένην οἰκίαν ἐν ᾗ καὶ αὐτῇ βιβλιοθήκη τις ἦν, πολλῶν μὲν καὶ ἄλλων βιβλίων μεστή ...
>
> But now the books on the Palatine perished on the same day as my own, when the fire ravaged not only the repositories by the Sacred Way, but first those at the Temple of Peace, and later those on the Palatine and the so-called House of Tiberius, where there was also a library which was filled with many other books... (Galen, *Peri Alupias* 18)

At least three libraries are mentioned here: the Palatine library near Augustus' temple of Apollo, the library in the Templum Pacis which is also well-attested by Galen and others, and the Domus Tiberiana library (which may be related to the library attested elsewhere in Tiberius' Temple of the Deified Augustus, restored under Domitian as the Templum Novum). Galen's library confirms that this Domus Tiberiana library existed and was housed somewhere in the path of the fire. It is possible (though I think overall unlikely) that Galen also refers in subsequent corrupt passages of this text to a fourth imperial library at Antium, presumably in the imperial villa there.[35]

---

33 At the risk of circularity, I think that we could propose that this market effect also worked the other way around: the clustering of readers provided an audience which encouraged further acts of imperial library benefaction. A similar effect is visible in comparing the clustering in time of imperial library benefaction in Rome and by private euergetistic donors in the cities of the eastern provinces of the empire in the early second century CE: who was imitating whom?

34 For the text: Boudon-Millot *et al.* (2010).

35 Nicholls (2014).

Galen's account of these libraries gives us a more detailed understanding of their operation and contents than we have seen before,[36] and presents their contents (and particularly reading and making copies of them) as more fundamentally important to his work than does (for example) his contemporary Gellius, who tends to use them as settings for discussion.[37] The picture that emerges is of a set of libraries that provided invaluable material to Galen as a literary scholar, his chief preoccupation in the *Peri Alupias*, because of the quality and number of the books they contained:

> Οὔτε οὖν ὅσα σπάνια καὶ ἀλ⟨λ⟩αχόθι μηδαμόθεν κείμενα δυνατόν ἐστιν εὑρεῖν [ἐστιν], οὔτε τῶν μέσων, διὰ δὲ τὴν τῆς γραφῆς ἀκρίβειαν ἐσπουδασμένων, Καλλίνια καὶ Ἀττικιανὰ καὶ Πεδουκίνια καὶ μὴν Ἀριστάρχεια οἵτινές εἰσιν Ὅμηροι δύο καὶ Πλάτων ὁ Παναιτίου καὶ ἄλλα πολλὰ τοιαῦτα, διασῳζομένων ἐντὸς τῶν γραμμάτων ἐκείνων αὐτῶν ἃ καθ' ἕκαστον βιβλίον ἢ ἔγραψαν ἢ ἀν⟨τ⟩εγράψαντο οἱ ἄνδρες ὧν ἦν ἐπώνυμα τὰ βιβλία. Καὶ γὰρ γραμματικῶν πολλῶν ἀντίγραφα βιβλία τῶν παλαῖων ἔκειντο καὶ ῥητόρων καὶ ἰατρῶν καὶ φιλοσόφων.
>
> And so it is impossible to find either those works that were rare and preserved nowhere else, or copies of standard works that were prized because of the precision of their text, those of Callinus, Atticus, Peducaeus and even Aristarchus, including the two Homers, and the Plato of Panaetius and many others of that sort, since preserved inside (the libraries) in each book were the words either written by or copied for the individuals after whom the books are named. There were also many copies of ancient grammarians, orators, doctors and philosophers. (Galen, *Peri Alupias* 13)

These libraries, then, were well stocked, reasonably well run, important resources for literary-minded individuals in Antonine Rome. Richard Marshall's paper in this volume shows us another example of a scholar in this landscape of people and libraries, in this case Suetonius (at one point in charge of the imperial libraries as *procurator bibliothecarum*). Marshall shows us Suetonius moving between a series of distinct library institutions in physically separate premises to compile his lives with their 'biblio-biographies': the same sense of movement through urban space between collections of books, and the potential of human actors to interrogate and supplement textual knowledge, that emerged from the anecdotal evidence discussed above.[38] Myrto Hatzimichali's chapter in this volume shows us the potential value of Rome's libraries for scholars who wished to consult works inconvenient for reproduction and wide circulation, such as Pamphilus' 95-book Lexicon.[39]

All of these libraries were situated in a relatively compact area near the Forum valley and up on the Palatine Hill. Readers and writers like these, on their way to

---

36 See, for example, Tucci (2008); Jones (2009); Nicholls (2010); Nicholls (2011). For Roman libraries generally and in particular their book collections, see Houston (2014).

37 Libraries as backdrop for learned *otium* and discussion in Gellius: the Domus Tiberiana (Gellius, *NA* 13.20.1); the Templum Pacis (*NA* 5.21.9, 16.8.2); the Forum of Trajan (*NA* 11.17.1). cf. Holford-Strevens/Vardi (2004) and Gunderson (2009).

38 See Marshall's chapter 'Suetonius the bibliographer; in the present volume, pp.119 – 145, and particularly its conclusions.

39 See pp.39 – 40 of the present volume.

and from the libraries, would have passed through the streets of booksellers discussed earlier,[40] providing a flow of potential customers which drove both the clustering of the shops and their address to passing pedestrian traffic.

This central part of town, where the emperor, magistrates, senate and courts transacted their business, naturally also contained archives. Many of these were apparently co-located with library spaces or nearby; indeed, Galen tells us that the warehouse on the upper Via Sacra where he stored his own book collection, for easy access to the libraries, also housed the official archives of four imperial procurators.[41] The enigmatic Roman state archive, the Tabularium, seems to have been located somewhere at the point where the Forum met the Capitoline,[42] and nearby the Atrium Libertatis, which also originally stood somewhere to the north east of the Forum on the shoulder of the Capitoline Hill, combined in Asinius Pollio's magnificent Augustan-era rebuild the character of both an archive and Rome's first public literary library; it may later have been replaced by or subsumed into the libraries of Trajan's Forum,[43] a complex which housed state archives in or alongside its libraries[44] and was occasionally the backdrop for staged imperial interventions in the official memory of the state, like the burning of public records of debt.[45] Some of the scholarly readers in these spaces had use for, or came across, archival material in their reading, and we might therefore add archives to the range of 'bookish places' in this area.[46]

The third category of purpose-built public cultural space in this region of Rome included the halls for performance, debate, lecture or recitation that often seem to have been co-located with Roman public libraries. We encountered in Gellius the idea that libraries themselves could be spaces for discussion as well as reading.[47] Galen extends this function to the entire Templum Pacis complex (which contained a library, and sat immediately to the south of the Vicus Sandaliarius), describing it as Rome's 'general meeting place for all those engaged in the learned arts, even before

40 Routes to the Forum, imperial fora and Palatine areas from the densely populated eastern suburbs of Rome in the Esquiline and Suburra regions funnelled traffic through this area.

41 Galen, *Peri Alupias* 8.

42 For a discussion of evidence for the location of Tabularium and Atrium Libertatis, see Purcell (1993); Tucci (2005); Coarelli (2010).

43 For the archival function of the Atrium Libertatis, see Livy, *Hist.* 43.16.13, 45.15.5. For Asinius Pollio, see Suetonius, *Aug.* 29.5, Pliny, *HN* 7.115; 35.23–25, 33–34. See *FUR* 28 for the presence of an [*ATRIVM*] LIBERTATIS in the Forum of Trajan.

44 E.g., SHA *Aurelian* 1.7, *Tacitus* 8.1–2 for the *libri lintei*; Gellius, *NA* 11.17.1 for Praetorian edicts.

45 SHA *Hadrian* 7.6, *Aurelian* 39.3.

46 E.g., the dubious SHA references in note 45 above, or, more reliably, Gellius, *NA* 11.17. Compare Coqueugniot's chapter in the present volume on the use of archival resources and compilations by Greek historians.

47 Gellius: Templum Pacis library (*NA* 5.21.9, 16.8.2); Domus Tiberiana (*NA* 13.20.1); Trajan's Forum (*NA* 11.17.1).

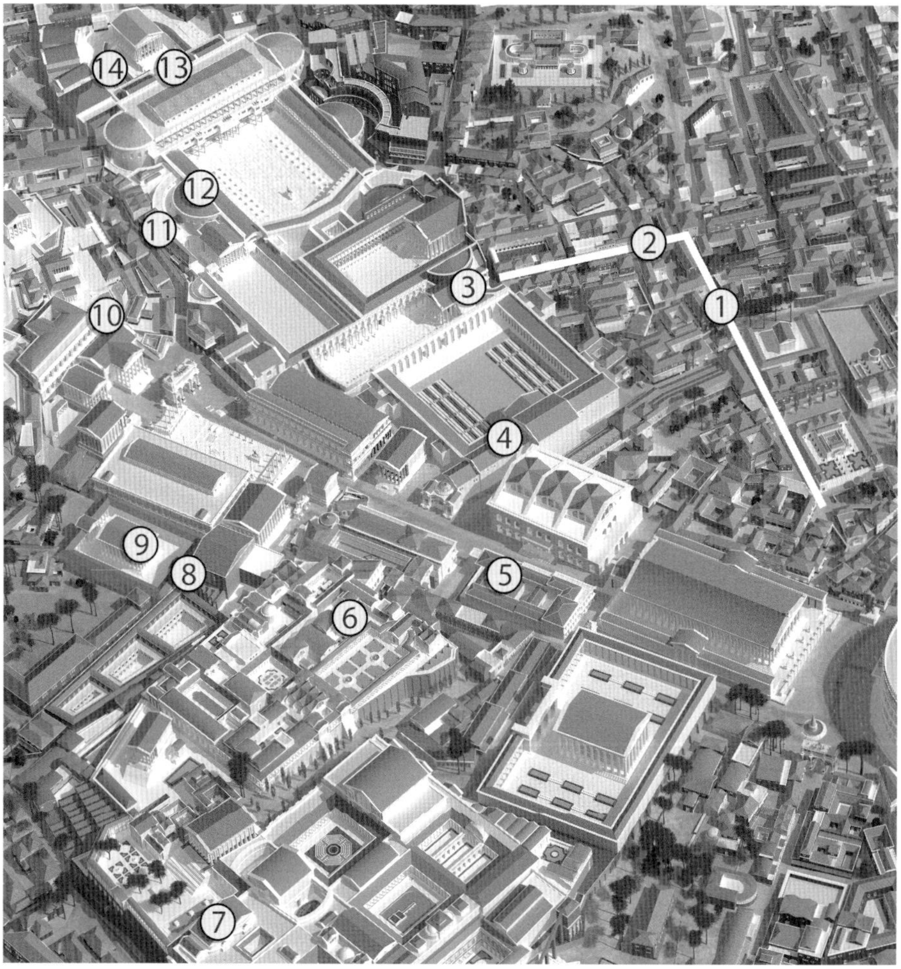

Approximate locations of places and buildings mentioned in the text. Image from author's digital model of ancient Rome c AD315; most buildings mentioned here (with the exception of Galen's destroyed warehouse) are still present. © Matthew Nicholls 2017

1. Route of Vicus Sandaliarius – **2.** Route of Argiletum – **3.** Porticus Absidata and route through Forum Transitorium to Forum – **4.** Templum Pacis – approximate location of library (and auditorium?) – **5.** Approximate location of Galen's warehouse – **6.** Approximate location of Domus Tiberiana – **7.** Palatine libraries – **8.** Vicus Tuscus – **9.** Possible location of Temple of Augustus – **10.** Approximate location of Tabularium – **11.** Basilica Argentaria – **12.** [ATRIVM] LIBERTATIS marked on marble plan fr.29 – **13.** Libraries of Trajan's Forum – **14.** Metro C audience halls ('Athenaeum')

Figure 4.1 Map of Bookish Rome

the fire.'[48] He also implies that a set-piece medical disputation in which he confronted his critics by producing huge armfuls of books took place in 'one of the great auditoria' of this complex, suggesting not only a general space for discussion and debate, but purpose-built formal rooms, presumably somewhere at the poorly understood southern end of the complex.[49] An inscription from near the second-century library of Celsus at Ephesus records the later construction of a pavement outside that building and a lost 'auditorium' or audience hall, implying a similar co-location of space for book-based and rhetorical or performative activity.[50]

Most significantly, excavations since 2006 for the new Metro C station at Piazza Venezia have uncovered the remains of a suite of three monumental marble-lined halls laid out in an arc, facing on to a piazza at the northern end of the Forum of Trajan. These three buildings, which can be dated precisely by quantities of brick-stamps to the years 123 and 125, were apparently largely identical in form and size. Each comprised a large rectangular hall containing two opposed banks of seating, entered at both ends from a wide lateral corridor. The main chamber had a concrete barrel vault. Although much despoiled by later re-use as copper-smithing workshops, then for burials in the 8th century, the internal decoration can be fairly confidently reconstructed. The floors of rectangular granite slabs with a giallo antico border is very similar to those of the two halls located to either side of Trajan's Column and conventionally identified as libraries.[51] The impression is of a complex of contemporary buildings in a prestigious imperial space, adorned with a consistent and monumental decorative scheme.[52]

It is tempting to identify these spaces with the lost Hadrianic Athenaeum, which is known to have combined library space with purpose-built auditoria, and to have been populated by learned audiences who sound similar to the crowds we have already encountered in libraries and bookshops.[53] Excavators have not yet found any

---

48 Galen, *Libr. Propr.* 2.21: καθ'ὅ τι καὶ πρὸ τοῦ καυθῆναι πᾶσιν ἦν ἔθος ἀθροίζεσθαι τοις τὰς λογικὰς τέχνας μεταχειριζομένοις.

49 Galen, *Libr. Propr.* 2.21: ... κατά τι τῶν μεγάλων ἀκουστηρίων... For the possible location of the Templum Pacis library, see Tucci (2013).

50 *Jahreshefte des Österreichischen Archäologischen Instituts*, VII (1904) Beibl. Col. 52; Platthy (1968) 157 no. 131. Cf. Burrell (2009).

51 For Trajan's libraries, see e.g., Packer (1997). Contrast Claridge (2007), which questions the conventional identification of these halls as libraries and proposed that they were 'auditoria-*cum*-honorary statue galleries', p.83. As Claridge agrees that a library or libraries somewhere nearby is firmly attested, this does not materially weaken the overall proposition here that libraries and auditoria were closely co-located within this part of Rome. Cf. Claridge's possible identification of a similar archive-library-auditoria cluster at the Pantheon, on p.79.

52 Claridge (2013) 5–18.

53 Athenaeum: SHA *Pert.* 11.3; *Alex. Sev.* 35.2; *Gordian* 3.4. Used for meeting of the senate: Dio 74.17. Judicial gatherings: Symmachus, *Ep.* 9.89.2. Libraries (or at least plentiful books): Sidonius, *Ep.* 2.9.4, *huc libri adfatim in promptu (videre te crederes aut grammaticales pluteos aut Athenaei cuneos aut armaria exstructa bybliopolarum)*. 'Here were books in abundance ready to your hand; you might have imagined yourself among the shelves of some grammarian, or the tiers of the Athenaeum, or a book-inscription giving a definitive identification, but we can certainly reconstruct this site as an important Hadrianic complex of audience or debate halls, in close proximity to the attested location of major imperial libraries. This looks like the firmest and most extensive evidence yet of the deliberate imperial provision of major facilities for orality, debate, recitation, or lectures, in close conjunction with library spaces laid out for the storage and consultation of written texts.

It was complexes like these, with their libraries, archives, and lecture halls that drew readers like Galen and Gellius through the nearby streets of booksellers with which we began, affecting the commercial character of the neighbourhood. Galen provides evidence of another effect of these imperial complexes on nearby commercial activity when he describes the loss of his possessions in a private, rented storage warehouse somewhere on the upper Via Sacra that was consumed in the fire of 192 CE, causing the loss of silver, gold, documents, medical instruments, and books.[54] Galen was not alone in renting space here, nor was this uniquely the practice of medical scholars: the *Peri Alupias* tells us that a grammarian, named in the Vlatadon ms of the text as the otherwise unknown Philides, also rented space in the destroyed warehouse, and lost so many books that he died of grief, while others went around grief-stricken at their losses as if in mourning.[55] It therefore seems that storage in this area was attractive to intellectuals of various disciplines who, we are told, paid for their accommodation on a commercial basis. The warehouse was not particularly conveniently located for access to the residential accommodation of any of these scholarly renters; Galen tells us that it was not near any private house.[56] Its security and its convenient proximity to the libraries of this part of Rome seem to have been what attracted Galen (and presumably the others), as it allowed him to keep his working materials, including his own notes, pharmaceutical recipe collections, and book manuscripts, close to the libraries where he pursued his wide-ranging intellectual projects.[57]

Educational spaces in the area may have been attracted by the same logic. Though specialist buildings for schooling are largely unknown in the Roman world, the lectures, debates and discussions that we saw in the porticoes, exedrae and lecture halls of the imperial fora formed the backdrop for the intellectual formation of aspirant lawyers and doctors; Gellius also records the presence of a young

---

seller's towering cases'. Seating: Sidonius, *Ep.* 9.14.2, *dignus omnino, quem plausibilibus Roma foveret ulnis, quoque recitante crepitantis Athenaei subsellia cuneata quaterentur.* 'Verily you deserve, if ever man did, the encouragement of Rome's applauding hands, the thunder of the Athenaeum hailing you master, till the seats shake with the clamour through every tier.'

54 Galen, *Peri Alupias* 4–6.

55 Galen, *Peri Alupias* 7. The ms is corrupt and the name uncertain; Boudon-Millot *et al.* (2010) suggest Philippides.

56 Galen, *Peri Alupias* 8.

57 For a fuller treatment of Galen's choice of this storage space, and its useful proximity to libraries, see Nicholls (2011) esp. 125–30.

man, *adulescens quispiam*, among the group discussing a text in an imperial library.[58] Further down the educational ladder, graffiti (an abecedarium and fragmentary poetic quotations including the first line of the *Aeneid*) suggest a school in the Trajanic structure known as the Basilica Argentaria at the northwest corner of the Forum Iulium, close to Trajan's libraries and not far from where the bookshop streets debouched into the Forum area.[59] For another example of the co-existence of a library with educational space we can turn to the roughly contemporary Library of Pantainos in Athens. This library presumably contained the scholarly and educational book collection built up by Pantainos and his school-teacher father;[60] its architecture and the graffiti on its columns have suggested spaces for discussion and education, perhaps comparable to the educational activity once sustained in the Ptolemaion, directly across the road from the new library.[61] Back in our corner of Rome, Peter White observed that four of Gellius' bookshop anecdotes include the presence of *grammatici*, and suggests that since these men were not able to make public performative demonstrations of their ability in the same way as orators or poets, the commercial space of bookshops might have provided 'a microenvironment in which they could shine by advertising and applying their particular knowledge of books',[62] a venue in which they could effectively advertise themselves to potential clients and patrons.

This is another example of the blurring of commercial and quasi-public intellectual space and activity that gave this part of Rome a complex, learned flavour. Taken together this range of places and activities suggests an urban landscape fairly well populated by potential readers, writers, listeners, customers, and visitors to these different sorts of space, and for the different sorts of staff who serviced them; this range of what we might think of as 'bookish people' might in turn have some implication for our understanding of what the 'public' identity and character of these monumental imperial complexes meant in practice, and also for the developing scholarship on different modes Roman literacy. Recent discussions of ancient literacies generally avoid a binary division of the literate from the illiterate,[63] fruitfully complicating the picture by considering, for instance, interactions with non-literary texts; the in-

58 Gellius, *NA* 13.20

59 Dudley (1967) 122 = BCAR, 61 IIIff: ABCEDFGIL ... MANTVA ME GENVIT ... ARMA VIRVMQVE CANO TROIAE QVI PRMVS AB ORIS .. SORACTI ... HECTOR ... CAECILIVS .. CAECILIVS ALVMNVS ... SMIRINA VALE.

60 Thompson/Wycherley (1957) 150; Shear (1935) 330–32 = Platthy (1968) 112 no. 36.

61 The inscribed dedication (and excavation of the site) indicates a building with stoas and a peristyle as well as a library, implying a place for meeting and discussion, perhaps adapted from a pre-existing school building. Thompson observes 'youthful' graffiti on the columns of the library's portico and deduces the presence of numerous young readers or pupils: 'the inference is that the Library of Pantainos is an extension or replacement of the library known to have existed in the Gymnasium of Ptolemy.' Thompson (1966).

62 *Grammatici* looking for business: White (2009) 285.

63 See, for example, the various chapters in Johnson/Parker (2009) or Frampton (2019).

tersection and interdependence of traditionally opposed categories like orality and literacy, even in a highly literate and book-centred age; recitation and private reading; the relatively widespread attestation of non-elite literacy such as that required for banking, commerce, and participation in democratic institutions, such and name literacy and list literacy; and the materiality of texts and authorial responses to it, connected to ideas of longevity and preservation. The busy population of human readers and authors considered above, ranging from school children to slave copyists to the emperor's physician, interacting in various ways with written texts and with each other, in private commercial and street spaces and among the institutional libraries and lecture halls of central Rome, is an interesting contribution to the study of this lively lettered world.

Serena Ammirati

Towards a Typology of the Ancient Latin Legal Book

Texts of legal content have often proved to be a fertile topic of research for scholars devoted to the history of ancient and medieval Latin book. Being an example of a specific textual genre (particularly with regard to various legal cultures and historical contexts), it is also distinctive from a codicological and palaeographical point of view. As I was able to ascertain in recent inquiries into the palaeography of the ancient Latin book,[1] there is also a quantitative factor which should be taken into consideration: in the period between the 1st and 7th centuries CE, nearly half of Latin written fragments of archaeological provenance are of legal content. This allows us to investigate the extremely detailed material and morphological history of the Latin book *tout-court*. A new appreciation of this documentation from a philological, palaeographical and legal perspective is one of the main goals of the REDHIS project, in which I am currently involved. The survey undertaken so far has led to some important results:[2] it has revealed how complex the transmission of the Latin legal knowledge must have been, and how multifaceted the morphology of its 'material carriers' was.[3]

My aim in this paper is to survey the earliest evidence of the ancient Latin book in order to determine the possible form (or forms) of such books.[4] Do these books have graphic and/or physical features in common? Was a legal book immediately distinguishable from a book of a different content? Papyrological and epigraphic evidence will be taken into consideration alongside literary sources. The present inquiry will be limited to texts up to and including the 3rd century CE, as the 4th century marked a significant watershed in the history of the ancient book. As is well

These findings are the result of research that has received funding from the European Research Council under the European Union's Seventh Framework Programme (FP7/2007–2013)/ERC grant agreement nr. 341102 'Redhis. Rediscovering the hidden structure. A new appreciation of Juristic texts and Patterns of thought in Late Antiquity', Università degli Studi di Pavia (Dipartimento di Giurisprudenza), Principal Investigator Dario Mantovani, Senior Staff Luigi Pellecchi (<http://redhis.unipv.it>).

1 Ammirati (2015a); Ammirati (2017).

2 A brief presentation of the project and its encouraging results can be read in Mantovani (2017); on the necessity of this approach for the history of Roman law, see also Mantovani (2018).

3 Cavallo (2002), 25–29.

4 By the phrase Latin 'legal book' I mean a book of a legal content, i.e., the work of a Roman jurist, either complete or anthologised, a collection of laws/imperial dispositions, or a commentary on a legal work/works or law/collection of laws. A full analysis of the textual genres can be found in Mantovani (2017) and (2018).

Serena Ammirati, Università degli Studi Roma Tre

https://doi.org/10.1515/9783110660982-006

known, the book in codex form, already attested and in use in previous centuries, particularly in the Latin world,[5] superseded the book roll and this, as has been noted, caused significant changes in the layout and distribution of texts in manuscripts.[6] The new book in codex form, in fact, offered new bibliological and graphic solutions. There was a denser concentration of text, the possibility to leave wide margins for annotations and commentaries,[7] and a different distribution of indices and organisation of topics and their sequence. As for the legal codex, it was developed and implemented largely under Theodosius II and Justinian: new customised scripts were introduced, as well as new formats in order to make the legal book recognisable not only by its content, but also by its morphology.[8] Some recourse to Late Antique evidence will be necessary to show how different layout and graphic changes might have occurred, particularly within the same text.

What the Sources Tell Us: *capita* and *rubricae* in literary texts, papyri, and inscriptions

In his letter to Trebatius (Cicero, *Fam.* 7.22), Cicero recalls a recent conversation about a possible legal controversy. To find a solution, he looks in his copy of a book by Q. Mucius Scaevola, where he discovers a pertinent reference; he copies it down and includes it in his letter. It is worth noting that he refers to the passage he wants to use as a *caput*.[9] The use of the word *caput* is interesting, as it suggests that the text (perhaps the *De iure civili*, in 18 books) was written and set down in columns in the rolls (with possibly one roll per book)[10] following a certain structure (i.e., sequence of topics) and divided into paragraphs, the borders of which must have been highlighted in some way. We do not know exactly what the book by Scaevola in Cicero's possession

5 A brief summary of the *status quaestionis* can be found in Ammirati (2013) 9–15.

6 In the conference, Zachary Cole offered a detailed discussion of the codex in Greek literature, particularly in Christian communities.

7 On this topic, see especially McNamee (2007).

8 As already pointed out in Ammirati (2010). The results and discoveries made under the auspices of the REDHIS project seem to strengthen the validity of this claim.

9 *Illuseras heri inter scyphos, quod dixeram controversiam esse, possetne heres, quod furtum antea factum esset, furti recte agere. itaque, etsi domum bene potus seroque redieram, tamen id caput ubi haec controversia est notavi et descriptum tibi misi, ut scires id quod tu neminem sensisse dicebas Sex. Aelium, M'. Manilium, M. Brutum sensisse. ego tamen Scaevolae et Testae adsentior* (translation Shackleton Bailey, Loeb: 'You made game of me yesterday over cups for saying that it was a moot point whether an heir can properly take action for theft in respect of a theft previously committed. So when I got home, though late and well in tipple, I noted the relevant section and send you a transcript. You will find that the view which, according to you, has never been held by anybody was in fact held by Sex. Aelius, Manius Manilius, and M. Brutus. However, for my part I agree with Scaevola and Testa'), Cicero, *Fam.* 7.22.

10 Wibier (2014) 56–72.

looked like, or how one *capita* was separated from the other, but we can imagine that the division of text into *capita* would have been familiar to the Roman writer. Indeed, a possible layout might have been the one found on the oldest papyrus fragment of a text by Cicero (*P.Iand.* V 90), *In Verrem* 2.3 – 4, which was on a roll. Written in ancient Roman cursive, it can be dated on palaeographical grounds to the 1st century CE: words are separated by *interpuncta*, a common feature in the oldest Latin fragments,[11] and sentences are divided by a large *K*, which has generally been interpreted as an abbreviation for *k(aput)*/*k(apita)*.[12] This idea is consistent with the sense of the text (where the *K*s do indeed separate small 'units of sense'), although it must be admitted that the length of each *caput* in *P.Iand.* V 90 is very short.[13] What we can reasonably infer from Cicero's letter to Trebatius, therefore, is simply that he possessed a copy of a legal book, the text of which was perhaps organised in *capita* that might have been visually distinguishable (how easily, we do not exactly know) and copied.

This latter feature is found in oldest Latin literary papyrological fragments, such as *P.Berol.* inv. 8507r, a copy of a speech that the emperor Claudius gave in the Senate in the year 47/48 CE on judicial reform. It is written in old Roman cursive and bears *interpuncta*. Three columns are preserved, the middle one in its entirety.

The papyrus is divided into paragraphs,[14] which are made evident by the protrusion of the first word/words of each section of the text (i. e., *ekthesis*) into the left margin of the column. This can be seen in col. II.2, *hae*, and col. III.10, *haec*. In col. II.1, the text ends before the margin limit of the column. This layout brings our discussion forward, proving that the text needed to be copied this way. A similar situation might have occurred for col. III.9, suggesting that col. I.8 must also have started with a projecting word, for only a minimum part of line 7 is written.

It is remarkable that the same structure occurs in a similar text relating to the same emperor, which is engraved on an extremely famous bronze tablet now kept in Lyon: the 'Tabula Lugdunensis' (*CIL* XIII 1668 = *ILS* 212). This tablet contains a speech by Claudius, again from the year 47/48 CE, where he, as censor, supports Gallic nobles in their wish to be accepted into the Senate. The speech is also reported by Tacitus (*An.* 11.23 – 25), and a comparison between the two versions of the text reveals

11 Ammirati (2015b) 13 – 15.

12 To be compared with the later PSI II 142, a papyrus leaf with a poetic reelaboration of Virgilian content, where verses are not written line per line, but continuously, separated by *K* or with the *H* which separates the epigrams in the famous papyrus of Cornelius Gallus (1st century CE, see Ammirati [2015a] 26, 63 and earlier bibliography). The latter is sometimes read as an abbreviation of *H(iatus)*.

13 Fioretti (2016). It has also recently been argued (Butler [2008 – 2009]) that the *capitulatio* (a word that is not attested in Latin, being a re-translation of the English word 'capitulation' and meaning, specifically, the division of a text into sections marked by each one beginning on a new line protruding to the left) that can be found in the oldest fragments of Cicero might refer back to the author himself. This, however, cannot be proved.

14 The full image is available at 'Berliner Papyrusdatenbank', http://berlpap.smb.museum/record/?result=0&Alle=8507, accessed February 14, 2019.

how the latter worked. The history of the request and its partial concession through a *senatusconsultus* only to the Aedui 'par considération de l'amitié qui les liait aux Romains'[15] is well known, as is the related scholarly debate. What concerns us here, however, is the actual layout of the text: letters are engraved in a beautiful capital script that is regular in size; words are separated by *interpuncta* and each paragraph, as in *P.Berol.* inv. 8507r, begins with a word/a group of letters engraved in the same size protruding into the margins. This disposition of the text is common in official public epigraphy (e.g., the *Fasti Capitolini*).

Epigraphic sources are a valid counterpart to those that evaluate books for they reveal striking similarities in solutions and layout. It has been commonly accepted, and rightly so, that legal texts engraved on stone or metal must have been based on an earlier copy on different writing material, possibly a papyrus roll. This model might also have contained instructions for the layout and proportions of the text, and the shape of the engraved letters also occasionally reveals that there was an antigraph on a 'softer' writing surface.[16]

As for 'layout strategies',[17] it is impossible in this time period to deduce whether changes in letter size based on the different lines/sections of the texts, which was very common in the epigraphy, were also a feature of books, although this seems highly unlikely.[18] There are other occurrences, however. Noteworthy examples are provided by the three Spanish *leges municipales*—Irnitana, Malacitana, and Salpensana—,[19] which contain, with atypical local adjustments, regulations for the civic life of a Roman *municipium* of the 1st century CE.[20] The three groups of bronze tablets contain text that partially overlaps. They are written in capital letters in different sizes, and are not in the same hand. It is important to note that in each of the three tablets, the *lex* is divided into smaller sections according to the different topics introduced by rubrics. Rubrics are preceded by an *R* crossed by a horizontal stroke under the bow, which stretches leftwards in the Malacitana and the Irnitana tablets. Here the text is indented and slightly aligned to the ideal left margin, which becomes more evident when the rubric consists of two or more lines. The first line of each paragraph begins with a word that protrudes into the left margin and we find the numbering of each paragraph/rubric under the protruding word in the Malacitana and Salpensana tablets. In the Lex Salpensana and Lex Irnitana, the letters are the same size; in the Lex Malacitana, where *ekthesis* is less pronounced, the first let-

15 Badoud (2002) 170.

16 See the examples collected in Edmondson (2014) 117–22, with accompanying bibliography; on the relation between bronze inscriptions and their antigraphs, see Mallon (1952) and Mallon (1982).

17 Fioretti (2012).

18 Decorte (2015).

19 González (1986).

20 The Leges Malacitana and Salpensana are now held in the Archaeological Museum in Madrid, while the Lex Irnitana is in the Archaeological Museum in Seville. For the texts, see D'Ors (1986) and Lamberti (1993).

ter is larger than the others. I believe this could also have been the original layout of a book collection of laws that was intended to be published and circulated.

More details about the materiality of the ancient legal book emerge from literary sources, where certain elements of the legal book appear to have become paradigmatic. A passage from Petronius' *Satyricon* has been often quoted to prove that legal books contained *rubricae*, i.e., headings/titles possibly (but not necessarily) written in red ink, as their name implies.

> *Emi ergo nunc puero aliquot libra rubricata, quia volo illum ad domusionem aliquid de iure gustare. Habet haec res panem.*
>
> I bought the child some books with red-letter headings in them a little time ago. I want him to have a smack of law in order to manage the property. Law has bread and butter in it.' (Petronius, *Sat.* 46)[21]

Petronius' text is interesting for several reasons. The character speaking, a guest of Trimalchio, is semi-illiterate, but the sentence is well constructed by Petronius, and the vocabulary emphasises material aspects: the approach to legal studies is referred to as *de iure gustare*, with an evident calembour between *ius*- 'law' and *ius*- 'soup' and its reward is *panem*. He uses the neuter *libra rubricata* rather than the correct masculine form, which seems to increase the vagueness of the category to which he is referring, and gives a peculiar nuance to the coeval *communis opinio*—which he seems to share—that studying law might lead to a rewarding career (*habet haec res panem*). The text also reveals that rubrics are an intrinsic feature of a book of legal content; it is notable that the speaker adds *aliquid de iure gustare* without any further explanation. This suggests that rubrics were essential parts of a legal book, but did not help in distinguishing between different genres of legal books and texts.

Two lines from Persius, *Sat.* 5.89 – 90 reveal that the word *rubrica* was also frequently used as a synecdoche for a legal book: *cur mihi non liceat, iussit quodcumque uoluntas excepto siquid Masuri rubrica uetabit?* ('Why am I not free to do everything that I want to do, excepting only what the red-titled Law of Masurius forbids?').[22] Moreover, in a wider sense, the word *rubrica* could indicate a legal text, but not necessarily, nor strictly, a legal book. In my opinion, this is what can be inferred by reading Quintilian, *Inst.* 12.3.11, *quorum alii se ad album ac rubricas transtulerunt et formularii vel, ut Cicero ait, legulei quidam esse maluerunt, tamquam utiliora eligentes ea quorum solam facilitatem sequebantur* ('Some of these went over to the praetor's album and the rubrics, and chose to be formularists or legal hacks'),[23] and Juvenal, *Sat.* 14.190 – 193, *clamosus iuuenem pater excitat: 'accipe ceras scribe, puer, uigila, causas age, perlege rubras maiorum leges; aut uitem posce libello'* ('Awake, boy,

21 Translation Michael Heseltine, Loeb.

22 Translation Susanna Morton Braund, Loeb. See Kissel (1990) 663, who quotes the relative scholium *rubrica vocat minium qui tituli legum annotabantur.*

23 Translation Donald A. Russell, Loeb.

and take your tablets; scribble away and get up your cases; read through the red-lettered laws of our forefathers ...').[24]

In Quintilian the coupling of *album ac rubricas* is notable in that it distinguishes the praetor's album from the rubrics of the laws (also with a very effective chromatic contraposition) and in Juvenal, red (*rubrae*) are the *leges* of the ancestors, which a *puer* must read thoroughly. As in Petronius, legal knowledge is once again paired with education. The connection between the rubrics and legal texts continues in Late Antiquity, where we find a line from Prudentius mentioning something which is prohibited by a rubrica: *Quae quia constituunt, dicant cur condita sit lex | bis sex in tabulis aut cur rubrica minetur | quae prohibet peccare reos...* ('but since this is what they lay down, let them tell us why law was established on the Twelve Tablets, why a statute which forbids wrong doing holds out its threats..., *Contra Symmachum*, 2.461–463).[25]

So *rubricae*, written in red or not, were only found in legal books and texts, whereas red ink was commonly used in Roman writing: on the epigraphical side, we have many painted inscriptions in red, and some evidence showing that red was also used to fill in the engraved letters.[26] In books, red was generally used to write the part of a text which needed to be highlighted (for example, to mark *indices*, *tituli*), as Cicero, Martial, Ovid, and Pliny the Elder relate.[27]

The scant papyrological evidence at our disposal seems to confirm this provisional picture: out of the two fragments of Latin *volumina* that contain red ink, the former, *P.Hamb.* II 167, is a fragment of a dramatic dialogue written in a beautiful capital script and dated to the 2nd century CE,[28] where the names of the characters who speak are written in red so as to highlight the different parts of this literary work. The latter is the oldest Latin literary papyrus of legal content, *P.Mich.* VII 456 + *P.Yale* inv. 1158r; it was written in ancient Roman cursive in the second half of the 1st century CE.[29] Besides *interpuncta* and a few 'epigraphic' abbreviations,[30] ll. 6 and 15 are noteworthy as they contain text in red ink written by the same hand and in the same writing style. This means that the text was organised into paragraphs, of which lines 6 and 15 were titles/headings, i.e., possibly *rubricae*.

24 Translation Susanna Morton Braund, Loeb.

25 Translation H. J. Thomson, Loeb.

26 Edmondson (2014) 127.

27 Dorandi (2007) 88.

28 Dingel (1973).

29 Ammirati (2015a) 28, 114.

30 See the abbreviation *d·f·o·* for *(d)are f(acere) o(portere)* in line 9, which also occurs in the text of the *Lex Gallia Cisalpina* from the year 41/42 CE, *CIL* XI 1146.

Rubrics from Antiquity to Late Antiquity

It is important to note that the same devices to mark the rubrics continued to be used in Late Antique manuscripts in codex form, with some minor differences. In *P.Vindob.* L 59+92 (V$^{in.}$ CE),[31] a codex fragment containing Marcian's *Institutiones*,[32] the rubric *De tutore honorario* is written in a different script than that of the main text (capital vs. early minuscule). It is slightly indented, and is marked by an *R* crossed by a diagonal stroke at the beginning and ending. The subsequent start of the line protrudes into the left margin; in the *Fragmenta Londiniensia Anteiustiniana* (V$^{in.}$ CE), possibly a copy of the Hermogenian Code,[33] the rubrics are centred, written in red ink in rustic capital letters (while the text is in uncial), proceeded by an *R* crossed by a diagonal stroke. According to extant evidence, it is not possible to determine whether an *R* also occurred at the end of the text, as it does in *P.Vindob.* L 59 + 92. In *PSI* XIII 1348, a collection of legal definitions of the 5th-6th centuries CE, a capital *R* crossed by a horizontal stroke occurs in the right-hand margin of a non-preserved fragmentary line, possibly with the same functions.

An alternative Solution for the Same Text: The *Institutes* of Gaius

Text layouts are not static and other formats could be adopted for copies intended for other uses. This is the case with a famous fragment of papyrus roll *P.Oxy.* XVII 2103,[34] which contains part of the fourth book of Gaius' *Institutes*, the most famous handbook of Roman law, to which Iustinian's *Institutes*, issued in the year 533 CE, are largely and explicitly indebted. Direct evidence of this work by Gaius is also known through the Gaius palimpsest manuscript in Verona (Verona, Biblioteca Capitolare, ms. XV 13), a parchment codex (6th century CE) that contains almost the entire work with some *lacunae* and difficulties in reading due to its poor state of conservation,[35] and some parchment leaves possibly found in Antinoe (*PSI* XI 1182),[36] datable to the end of the 5th century CE.[37] In all these three examples parts of the

---

31 A complete image is available on the Austrian National Library website, http://data.onb.ac.at/rec/RZ00008496, accessed August 12, 2017; see also Ammirati (2015a) 68–69.

32 Fressura/Mantovani (2018).

33 Corcoran/Salway (2012).

34 Hunt (1927) 173–90.

35 Ammirati (forthcoming).

36 The full image is available at 'PSI online', http://www.psi-online.it/documents/psi;11;1182, accessed 14 February 2019.

37 A re-edition of the text is currently being undertaken by Marco Fressura as part of the REDHIS project.

fourth book are transmitted, and this represents an unique occasion to compare layout and graphic strategies for the same text in antiquity and late antiquity, revealing that choices in text's division could not be all referred to the author himself.

P.Oxy. XVII 2103 consists of two pieces, the larger still showing the remains of three columns of writing. The script, dated to the 3rd century CE, is an ancient cursive written by an expert hand, which seems to be quite accustomed to documentary writing. The papyrus was found with other Greek documentary texts and a Latin literary historical fragment (*P.Oxy.* XVII 2088), all relating to Sarapion, alias Apollonianus, a wealthy man from a very influential family attested in Arsinoites and Ermopolites in the 2nd-3rd centuries CE.[38] Some similarities are noticeable between the hand of *P.Oxy.* XVII 2103 and other texts found in the same 'Sarapion archive', supporting the hypothesis that this could have been a private copy written in a more current hand and not meant for publication.

Besides the script, a few other material characteristics are noteworthy: the middle column bears a number in the upper margin (i.e., XVIIII), possibly written by the same hand as that of the text. This means this column must have been the 19th in the roll to which it belonged. This is consistent with the idea that this roll might have contained only the fourth book of the *Institutes*, therefore demonstrating that the *Institutes* conform to the traditional ancient book division of 'one book / one roll'.

The text is written in *scriptio continua* with only a few interpuncts visible. Its division into *capita* is respected by the scribe, but instead of starting a new *capita* on a different line, he uses another method: a *paragraphos*, which is written in the left margin to mark the line where the passage from one *caput* to another occurs. A corresponding medial point is written where the exact shift occurs within the line. An example of this can be found in ll. 42, 52 (Gaius, *Inst.* 4.72). It is interesting to verify whether the same text distribution occurs in the same places in later manuscripts. In the Codex Veronensis there is no sign of division (f. 17v) in the first place, while the second is not legible because of the poor condition of the page (f. 73r); it is not possible to ascertain whether *P.Oxy.* XVII 2103 had rubrics. Nonetheless, it is worth noting that rubrics occur in the Codex Veronensis and in the Antinoe fragments, although they are not consistent. For example, in *PSI* XI 1182 on page D we see traces of a rubric in red ink in lines 2–3. However, in the corresponding part of the Codex Veronensis (f. 18v) there is no marked text passage. Moreover, another rubric is added by a second hand in *PSI* 1182, page D, line 16, that is not found in the other codex.

It is possible to argue, therefore, that some divisions in text could not be attributed to the author himself, and may be the result of various distributions adopted by later copyists and readers during subsequent stages of textual transmission.[39]

---

38 See especially Moioli (1987) and Houston (2007).

39 Mantovani (2015).

Palaeography

In the case of *P.Oxy.* XVII 2103, the script is an important clue in verifying the possible context and use of the fragment. This naturally poses the question: is script, like the rubrics, a feature which can distinguish a legal manuscript from a non-legal one? We can only give a partially affirmative reply, and only in the case of Late Antique manuscripts, especially those copied under Theodosius II and Justinian. For the earlier period under consideration here, evidence does not allow us to say that script might have been a distinctive feature of text typology. Nevertheless, it is still a significant element when identifying context and determining whether a copy was produced for publication or not. In the case of legal content, this can be very important: the presence of informal scripts reveals that legal texts were not only produced for publication and public archives, but possibly also circulated among individuals for various purposes.

I will provide a brief survey of the papyrological evidence, highlighting the issues raised on a case-by-case basis:

a) Based on our limited knowledge, I believe the first representative example is *P.Mich.* VII 431, a fragment of a papyrus leaf, which contains the same sequence repeated sixteen times in sixteen successive lines; since the sequence is *cum institua*'[-*mus?* this has been related to a legal context since its very first edition.[40] This supposition might be true, but it cannot be proved. What is almost certain, however, is the fact that this is a writing exercise, where somebody wrote the first line in a calligraphic capital script, and where thin and thick strokes were lined and alternated precisely; what follows is in a different type of script, where strokes are thinner, the *ductus* faster and 'chiaroscuro' not pursued. I believe this was meant as practise for the different types of majuscule that existed, and they were employed according to the graphic skills of the scribes and the texts to be copied.

b) *P.Heid.* L 3 is a fragment from a papyrus roll written in a beautiful capital script. *Interpuncta* are generally marked, and thick and thin strokes alternate. Although it might appear to be ancient, i.e., similar to scripts of the 1st and 2nd centuries CE, reference to an emperor of the Severian family means that it cannot be prior to the end of the 2nd century CE. Fara Nasti has recently re-edited the fragment and believes that it contains evidence of a disposition of Alexandrus Severus concerning the *donatio mortis causa*.[41] Although it remains impossible to ascertain the authorship of the text, it is quite likely that *P.Heid.* L 3 was a book copy.

c) The same cannot be claimed with any confidence for *P.Aberd.* 130 and *P.Monac. Lat.* 2, fragments of papyrus rolls both containing fragmentary columns of text. The two are believed to be of legal content, as some 'legal' words can be distin-

40 Welkenhuysen (1968) 309–16.

41 Nasti (2016) 113–22.

guished, and both are datable to the 2nd century CE. They evidence a similar majuscule script type with some cursive and minuscule elements. In both cases, there is not enough text to ascertain whether the legal content belongs to a document or to a legal work, although, given the lack of calligraphy, the scripts do not immediately suggest a 'luxury edition'.[42]

Latin Characters in Search of an Author

A notable example of a calligraphic legal book brings the present survey to a close. *P.Fay.* 10 + *P.Berol.* inv. 11533 a+b + London, British Library inv. 2585 is a group of three fragments of papyrus, possibly belonging to the same original scroll or related scrolls. So far, only *P.Fay.* 10 and *P.Lond.* inv. 2585 have been published,[43] the edition of the Berlin being in progress.[44] The *editio princeps* of *P.Fay.* reveals a textual coincidence with *Digest* 29.1.1, where a passage from the XLV book of Ulpian's *Ad edictum* is given. In this part of his work, Ulpian is quoting a *caput ex mandatis* of the emperor Trajan, regarding the *testamentum militis:*

> *Militibus liberam testamenti factionem primus quidem divus Iulius Caesar concessit: sed ea concessio temporalis erat. postea vero primus divus Titus dedit: post hoc Domitianus: postea divus Nerva plenissimam indulgentiam in milites contulit: eamque Traianus secutus est et exinde mandatis inseri coepit caput tale. caput ex mandatis: "Cum in notitiam meam prolatum sit subinde testamenta a commilitonibus relicta proferri, quae possint in controversiam deduci, si ad diligentiam legum revocentur et observantiam: secutus animi mei integritudinem erga optimos fidelissimosque commilitones simplicitati eorum consulendum existimavi, ut quoquomodo testati fuissent, rata esset eorum voluntas. faciant igitur testamenta quo modo volent, faciant quo modo poterint sufficiatque ad bonorum suorum divisionem faciendam nuda voluntas testatoris"*
>
> The deified Julius Caesar was in fact the first to concede unrestricted *testamenti factio* to soldiers; but that concession was temporary. However, later the deified Titus first gave it; after this Domitian; thereafter, the deified Nerva conferred the fullest indulgence to soldiers; and Trajan followed this, and thenceforth such a chapter came to be inserted in [*imperial*] mandates. A chapter from the mandates: 'As it has been submitted to my notice that wills left by our fellow soldiers, which could be open to dispute if regard were had to the diligent observance of the laws, are repeatedly submitted; following the openness of my heart toward those excellent and most faithful fellow soldiers, I thought that provision should be made for their inexperience [*in legal matters*], so that whatever the way in which they made their wills, their wishes should be confirmed. Therefore, let them make their wills in any way they wish, let them make them in any way they can, and let the bare wishes of the testator suffice to settle the distribution of their property'. (Justinian, *Digest* 29.1.1)[45]

42 Ammirati (2015a) 29, 31, 83.
43 Grenfell/Hunt (1900) 99–100.
44 Ammirati (2015a) 83, 84; Salati (2018).
45 Translation from Watson (1998) 867.

P.Fay. 10 contains exactly the same text as the *mandatum* quoted by Ulpian. In the papyrus, each paragraph begins with a letter of a larger size slightly protruding into the left margin. Between the *Digest* and the papyrus there is only a small divergence found in the last fragmentary line of the latter. It is impossible to determine from *P.Fay.* 10 alone whether the entire papyrus contained the text of Ulpian or the imperial *mandatum*. However, what we can read in both the Berlin and London fragments (which have the same layout, but with a text unknown elsewhere) leads us to surmise that the original scroll (or group of scrolls) must have contained the imperial dispositions rather than a juridical work.[46] Nonetheless, it is possible that Ulpian might have extrapolated the *caput* from a book that was similar in appearance and layout, beginning his quotation with the caput *Cum in notitiam meam...* We know for sure that the text in *P.Fay.* 10 + *P.Berol.* inv. 11533 a-b + London, British Library, inv. 2585 was organized *in capita*. In this way, smaller sections would be easy to track. Indeed, this leads us back to Cicero and the way he was able to find, copy, and transmit Sabinus' thoughts in his letter to Trebatius, possibly in a papyrus roll, and maintain a similarity in appearance and layout.

Further details need to be added to these considerations. In the earliest extant copy of the *Digest* (i.e., the famous Florentine codex of the Pandects, Florence, Biblioteca Medicea Laurenziana s.n., post 535 CE), the text of Ulpian is copied without any marking of the passage from the main text to the quotation, but the quotation is highlighted in the margin by a sequence of '>' signs running parallel to the text. However, we do not know what Ulpian's archetype looked like, which must have been from the same period as the papyrus. On palaeographical grounds, in fact, we can assign *P.Fay.* 10 + *P.Berol.* inv. 11533 + London, British Library, inv. nr. 2585 to the middle 3rd century CE, as the script is an extremely calligraphic ancient cursive, with some minuscule elements: the letter *B* has the typical ancient cursive form, with the 'panse à gauche';[47] *d* already has the minuscule form with the second stroke (which should be upright) completely leaning leftwards, thus closely resembling the later uncial form; *R* anticipates its later *B-R* uncial form, with its second descending stroke completely horizontal. Strokes are very thin and the letters are regular in size and slope slightly to the right. It is plausible that the original book might have been an official copy of imperial *mandata*.

It is interesting to compare this script to that of the coeval *P.Berol.* inv. 25674,[48] a legal fragment of documentary content, an edition of which is forthcoming. The same letter forms are found, yet, they are drawn in a more informal and current way. This

46 As is clear, for example, from the expression *fisco meo* in *P.Berol.* 11533b, l. 8.

47 The study of the evolution of the form of *B* from the ancient capital drawn with four strokes to the later cursive two-stroke form with 'panse à gauche' (looking like a minuscule *d*), and on to the fixed minuscule form it has today (*b*) was one of the main concerns of the French school of palaeography in the first half of the 20th century: see Mallon (1952) *passim*.

48 The *editio princeps* by Timothy Teeter will appear in the next volume of *Berliner Griechische Urkunden*.

comparison allows us to surmise that there may have been a fairly standard, regular sloping script (a mixture of capital, ancient cursive and early minuscule elements), which was used in the first half of the 3rd century CE both for documents and books. The script is more informal and rapid solutions for documents (more ligatures between letters, fewer serifs, and a greater tendency to slope), while those for books were more calligraphic.

Conclusion

This survey has aimed to show that for the early period (1st-3rd centuries CE) we cannot talk of precise, definite typologies of Latin books of legal content. Their scripts, formats and dimensions are perfectly consistent with what we know of other Latin books of different textual genres: various scripts, layouts and book formats were used during the same periods for the same texts (or types of texts). Regarding other textual genres, it is worth noting that those most commonly found are glossaries[49] and writing exercises, mostly of Virgilian content. This shows, even for this earlier period, that there was a wide range of layouts and graphic options.[50] A recognisable and definite book typology for the Latin book of legal content is not attested until the end of the 5th century CE, when a new customised script, the *B-R* uncial, starts to be regularly adopted, and codices present similarities in the size and layout of the texts.

However, the comparison between the papyrological, literary and epigraphic evidence has shown that some regularities already occur in the earliest items attested, such as the regular use of rubrics for legal texts only, and the distribution of text in *capita*, often highlighted by *ekthesis* and enlarged initial letters. As we have seen, the same format was adopted in legal codices, and so it is reasonable to assume that there must have been some continuity in style and layout. This is also consistent if we consider the history of the Latin legal book in the wider framework of the history of the ancient Latin book: we know that works such as those by Vitruvius and Pliny the Elder must have given rise to books with complex layout solutions to facilitate consultation, to contain drawings and so on, or that collections of poetry, such as epigrams, might have resulted in a particular, visually-identifiable layout. We see this, for example, in the papyrus of Cornelius Gallus, and in the coeval metrical inscriptions and graffiti from Pompeii and Rome.[51]

49 A recent detailed survey can be found in Dickey (2016) and Fressura (2017).

50 An attempt was made to survey all of them in Ammirati (2015a) 27–28, 36–37.

51 On the layout of the Gallus papyrus, see Capasso (2003). In the elegiac couplet, both verses have the initial letter which is larger in size, while the hexameter is slightly projected into the left margin; small poems are also separated by a sort of *H*, which has been interpreted as H(iatus) (see Capasso, *Il ritorno*). On the metrical graffiti of Pompeii and Rome, which largely have the same layout even when verses are not of different lengths (such as *CIL* IV 923, both pentameters, thus demonstrating there

In conclusion, therefore, even if is not possible to identify a precise ancient typology of Latin legal books, given the wide variety of scripts and formats it is legitimate to affirm that there must have been some regularities of layout that could have made such texts physically distinct from other books. The scant extant evidence does not allow us to say whether these distinct features were there from the beginning of Latin book production (we have no fragments earlier than the 1st century CE), but it is quite significant that, as we have seen, the earliest comparable evidence (epigraphic, papyrological, and literary) indicates a certain consistency, and that such regularity (e. g., the *rubricae*) was present through the centuries as the book form developed from roll to codex.

might have been a common idea of what a book of poetry should look like), see Cugusi (2007) and Cugusi (2016).

Stephanie Roussou

New Readings in the Text of Herodian

This paper focusses on a selection of significant new readings in the text of Pseudo-Arcadius' *Epitome* of Herodian's Περὶ καθολικῆς προσῳδίας, based on newly-discovered manuscripts, and discusses what they mean for our understanding of Herodian and his work. At the same time this paper tries to illustrate, where possible, the sources from which Herodian derived material as examples for his accentuation rules. The discussion of these new readings arises from my recent new critical edition of Pseudo-Arcadius' *Epitome*.

Herodian and his Περὶ καθολικῆς προσῳδίας

Herodian is the grammarian to whom we owe most of our knowledge of Greek accentuation. He lived in the second century CE and he was the son of another grammarian, Apollonius Dyscolus.[1] His Περὶ καθολικῆς προσῳδίας ('On Prosody in general') was the most important ancient work on Greek accentuation and the first systematic treatment of ancient Greek prosody to have a substantial and lasting impact on ancient and medieval Greek scholarship and teaching.[2] This work has not come down to us in its original form, but in epitomes and fragments.[3] The most important sources for this work are two epitomes, one misattributed to Arcadius (conventionally called Pseudo-Arcadius' *Epitome*) and another by John Philoponus of Alexandria.[4]

The sources for the Περὶ καθολικῆς προσῳδίας

The reconstruction of Herodian is not a straightforward issue, because the two epitomes differ significantly from each other. Philoponus deals with the accentuation of the oblique cases once the accent of the base or dictionary form of a word is known, whereas Pseudo-Arcadius' *Epitome* is mainly interested in which syllable of a base

---

1 See also Dyck (1993) esp. 772 and Dickey (2007) 75.

2 See also Vendryes (1945) 11; Dyck (1993) 776; Probert (2006) 22; and Dickey (2007) 75. For further discussion of Herodian, see the chapter by E. Dickey in this volume.

3 See also Dyck (1993) 776 – 82 and Dickey (2007) 75 – 76.

4 See also Dyck (1993) 776 – 77 and Dickey (2007) 76. On the authorship of Pseudo-Arcadius' Epitome, see Lentz (1867) cxxx-cxxxv; Egenolff (1887) 6; Galland (1882) 12 – 16; Cohn (1895) 1153 – 56; and more recently Kaster (1988) 244, 366 – 67, 385; and Roussou (2018), section 2.2.1. John Philoponus' Epitome has been edited by Xenis (2015a).

Stephanie Roussou, University of Cyprus

https://doi.org/10.1515/9783110660982-007

form will have the accent, on the basis of the word's termination, and examines the accentuation of nouns and adjectives in the nominative according to their termination in fourteen books, but the accentuation of the oblique cases in only one book.[5] In addition to the two fully-preserved epitomes, two fragmentary epitomes have been discovered in the 20th century, a fourth-century parchment codex from Antinoöpolis (LDAB 1117) and a palimpsest dated to the first half of the 10th century (Vienna, codex Vindobonensis Hist. gr. 10).[6] A comparison of the two fully-preserved epitomes with the two fragmentary epitomes shows that the arrangement of the material in Pseudo-Arcadius' *Epitome* shares similarities with the two fragments, and suggests that Pseudo-Arcadius retained the arrangement of the material as in Herodian's original work.[7]

Lentz's edition of Herodian's Περὶ καθολικῆς προσῳδίας

Lentz attempted a continuous reconstruction of Herodian's lost Περὶ καθολικῆς προσῳδίας on the basis of various Herodianic sources,[8] including texts which are likely to have derived material from Herodian either directly or indirectly. Lentz had some good ideas about how to combine these sources and reconstruct the text, and he made some good conjectures. Despite the contribution of Lentz's edition to the study of Herodian's lost Περὶ καθολικῆς προσῳδίας, one needs to be careful when using it, because in producing a continuous text Lentz mixes material from the two epitomes and from other texts thought to have derived material from Herodian, and he even invents his own rules and adds his own examples.[9] In addition, the attribution of some of the material to Herodian is not always secure.[10] When employing Lentz's edition one needs to consult his apparatus constantly to see which texts Lentz employs for his reconstruction of Herodian, and it is much more practical and methodologically correct to consult Lentz's sources directly when possible.[11]

5 See also Dyck (1993) 777.
6 See Hunger (1967) and Wouters (1979) 216 – 224, and more recently Dyck (1993) 779 – 82 and Dickey (2007) 76.
7 See also Roussou (2018) section 2.2.1.
8 See also Dyck (1993) 775 – 78 and Dickey (2007) 76.
9 See also Dyck (1993); Dickey (2007) 76; and Roussou (2018) section 3.
10 See Egenolff (1900); Egenolff (1902); Egenolff (1903); Dyck (1993) 776.
11 For a critique of Lentz's edition, see also Chandler (1881), xx; Dyck (1993) 775 – 78; Dickey (2007) 76.

The editions of Pseudo-Arcadius' *Epitome*

My new critical edition of Pseudo-Arcadius' *Epitome* is a response to the need for new and separate editions of the sources for the Περὶ καθολικῆς προσῳδίας, previously edited by Barker in 1820 and by Schmidt in 1860. Neither of these two editors consulted all the extant manuscripts, and they did not consult the manuscripts at first hand but through transcriptions prepared by other people.[12] Barker's edition is essentially a transcript of the least reliable manuscript (codex Parisinus Graecus 2102, C), accompanied by variant readings from its codex gemellus (codex Parisinus Graecus 2603, B).[13] Barker consulted this manuscript on the basis of a collation made by Gregorius Georgiades Zalykius.[14] This collation sometimes contains errors. It sometimes attributes to these manuscripts the wrong readings, and it often fails to note the variant readings of manuscript B. Both these manuscripts contain scribal errors and omissions, not only of single words but of whole accentuation rules.

Schmidt's edition was mainly based on three manuscripts, the two which Barker employed (B and C) plus another one, codex Hauniensis Regius 1965 (A). In fact, Schmidt did not collate these manuscripts himself, but he used B and C from Barker's edition and A on the basis of the variant readings noted by Bloch, who made the collation.[15] This collation contains some errors and fails to report some of A's variant readings, while at the same time Schmidt's edition inherits some of the mistakes of Barker's edition.[16] Moreover, in cases where some of A's variant readings were not reported in Bloch's collation, Schmidt printed Barker's text, which was mainly based on the least reliable manuscript, C, although this new manuscript could have offered a good reading. This is especially problematic because in many cases the reading of the least reliable manuscript (C) is different from the rest of the manuscripts, since its scribe intentionally made certain changes, either aiming to create a more complete text or a text closer to the classical style. Nevertheless, Schmidt managed to improve the text to a large extent on the basis of some of A's good readings, by making important conjectures of his own, and by adopting conjectures made by other scholars.

My edition is the first to take into account all the extant manuscripts that are known to us today. A study of Pseudo-Arcadius' manuscripts shows that codex Matritensis 4575 (M) and codex Baroccianus 179 (O), which are used for the first time in my edition, are the most important, and enable us to improve the text to a consider-

12 See Roussou (2018) section 5, with further bibliography.

13 Barker (1820) v.

14 Barker (1820) v.

15 Schmidt (1860) v. Bloch's collation was published by Dindorf (1823) 48–70.

16 See also Galland (1882) 4. Egenolff (1881) provides the results of his new collation of A and corrects Bloch's errors.

able degree.[17] This is especially important to the extent that some scribal errors printed in the two previous editions, can now be replaced by good readings.

The manuscript tradition of Pseudo-Arcadius' *Epitome*

The history of manuscripts M, O, and A is particularly interesting as they appear to have been copied in the same place, around the same time, and by scribes who knew each other. Manuscript M was written by Constantine Lascaris in Messina,[18] copied from a παλαιὰ βίβλος, as he himself states in codex Matritensis 4689 (139): μόλις εὗρον τὸ περὶ τόνων τοῦ Θεοδοσίου ἐν βίβλῳ παλαιᾷ τῆς μονῆς τοῦ Σωτῆρος τῆς ἐν ἀκρωτηρίῳ τῆς Μεσσήνης ('I only just found Theodosius' Περὶ τόνων in an old book in the monastery of the Saviour on the promontory of Messina'). It is dated to the 15th century, most likely no later than 1482, as Galland argues,[19] because in 1482 Lascaris had prepared an abbreviated version of the sixteenth book of the Περὶ καθολικῆς προσῳδίας for the use of Jacob Ximenes. Therefore, Lascaris must have already copied Pseudo-Arcadius' *Epitome* of the Περὶ καθολικῆς προσῳδίας, before excerpting the sixteenth book, since he must have had in his possession a copy of the epitome before he excerpted the specific book.

Manuscript O was written by Leon Chalkiopoulos (Λέων Χαλκιόπουλος) from Crete, in Messina in 1495.[20] We know the exact date because this manuscript has a colophon. Although we cannot be sure if Leon was a pupil of Constantine Lascaris, Leon's scribal activity in Messina in the last decade of the 15th century suggests that he must have known Lascaris.

Manuscript A was written by Urbano Valeriano, who was a pupil of Constantine Lascaris, the scribe of M.[21]

O is a contaminated manuscript deriving from M as well as M's source, the παλαιὰ βίβλος. There are some good readings in O which are not in M. These must have been inherited from the παλαιὰ βίβλος. The good readings in O alone are rare words and it is difficult to think that these could be owed to the scribe of O. On the other hand, there are some errors in O which are most probably due to certain ambiguous readings in M. Since it is not possible for us to know whether the primary source for O was M or the παλαιὰ βίβλος I have employed a dotted line to indicate the relationship between O and M, and O and the παλαιὰ βίβλος.

17 See further Roussou (2018) section 4.

18 On Lascaris' handwriting, see the discussions of Gamillscheg *et al.* (1981-) 1.223; Fernández Pomar (1966) esp. 235; and Martínez Manzano (1994) 307.

19 Galland (1882) 7–8.

20 The name of the scribe and the date are mentioned in the colophon. For further information on this scribe, see Gamillscheg *et al.* (1981-) 1.237.

21 Smith (1978) esp. 58–59; see also Schartau (1994) 175. This scribe is also called Urbano Bolzanio, or Bellunese or Urbano da Belluno. On this, see Adler (1938) esp. 77; Martínez Manzano (1994) 25 n.85.

A derives from M. This can be concluded from errors in A which are due to ambiguous letters in M. Sometimes A and O agree in error against the rest of the tradition, and this suggests that they both inherited those errors from the παλαιὰ βίβλος. The errors shared by A and O are errors that Lascaris could have corrected in his own copy.

Manuscripts B and C are *codices gemelli* deriving from π, which in its turn is a *codex gemellus* with manuscript A, deriving from M.

The stemma of the manuscript tradition of Pseudo-Arcadius' *Epitome* is as follows:

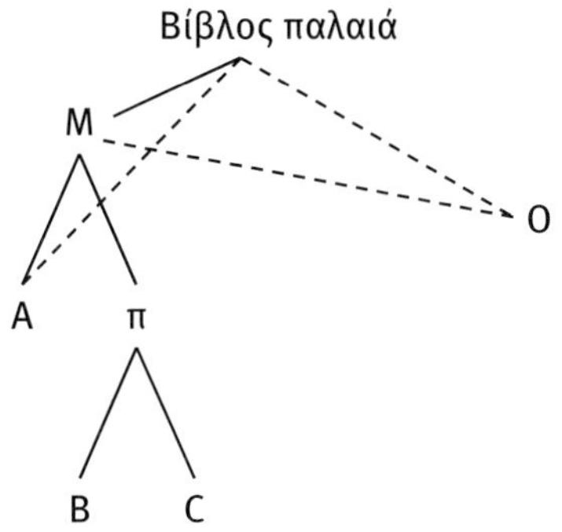

The arrangement of the material in the Περὶ καθολικῆς προσῳδίας

The Περὶ καθολικῆς προσῳδίας consisted of 20 Books: the first nineteen books treated systematically the rules for the accentuation of Greek on the basis of their terminations, while Book 20 dealt with breathings and vowel quantities.[22] The work also included an appendix on the accentuation of words in connected speech.[23] The arrangement of the material of the first nineteen books follows the same order for the eight parts of speech as the Τέχνη γραμματική attributed to Dionysius Thrax:[24] *GG* I.I 23.1–2 Τοῦ δὲ λόγου μέρη ἐστὶν ὀκτώ· ὄνομα, ῥῆμα, μετοχή, ἄρθρον, ἀντωνυ-

22 On the history and transmission of the so-called Book 20 in Pseudo-Arcadius' Epitome, see Roussou (2018) section 2.2.1.

23 For a discussion of the Appendix, sometimes referred to as Book 21, see Roussou (2018) section 2.2.1.

24 On the highly debated issue of the authenticity or not of this *Ars Grammatica* one may consult the essays in Law/Sluiter (1995), and the note by Wilson (2007) esp. 69–70, where the authenticity is defended.

μία, πρόθεσις, ἐπίρρημα, σύνδεσμος ('The parts of speech are eight: nominal[25], verb, participle, article, pronoun, preposition, adverb, conjunction').[26] In fact, both the Περὶ καθολικῆς προσῳδίας and the Τέχνη γραμματική follow the Alexandrian grammatical tradition in terms of the distinction of the parts of speech.[27] Books 1–15 deal with nominals, 16 and 17 with verbs and participles, 18 with so-called articles,[28] pronouns, and prepositions, and 19 with adverbs and conjunctions. The first 14 books treat the accent of the nominative singular forms according to their terminations, while Book 15 deals with the accent of the oblique cases. Books 1–10 treat the masculine and feminine nominals ending in consonants, 11 and 12 the feminine nominals ending in vowels, 13 the neuter nominals, and 14 the monosyllabic nominals. The masculine and feminine nominals in Books 1–12 are arranged in the same way as in the Τέχνη γραμματική (*GG* I.I 15.1–16.1) attributed to Dionysius Thrax, so that Books 1–10 treat the masculine and feminine nominals ending in ν, ξ, ρ, σ, ψ, Book 11 the feminine nominals ending in α, and Book 12 the feminine nominals ending in η and ω. The only difference between the two texts in the order in which the terminations are dealt with is that the Τέχνη deals with feminines with vowel terminations before those with consonant terminations, whereas in Pseudo-Arcadius' epitome feminines ending in a consonant have already been treated in the first ten Books together with masculines, and therefore only feminines ending in vowels are dealt with in Books 11 and 12:

> Τελικὰ ἀρσενικῶν ὀνομάτων † ἀνεπεκτάτων κατ' εὐθεῖαν καὶ ἑνικὴν πτῶσιν στοιχεῖά ἐστι πέντε· ν ξ ρ σ ψ, οἷον Δίων Φοῖνιξ Νέστωρ Πάρις Πέλοψ. θηλυκῶν δὲ ὀκτώ· α η ω ν ξ ρ σ ψ, οἷον Μοῦσα Ἑλένη Κλειώ χελιδών ἕλιξ μήτηρ Θέτις λαῖλαψ.
>
> The terminations of the masculine nouns that are parisyllabic in the nominative singular are five: ν ξ ρ σ ψ, for example Δίων Φοῖνιξ Νέστωρ Πάρις Πέλοψ. And (the terminations) of the feminine nouns are eight: α η ω ν ξ ρ σ ψ, for example Μοῦσα Ἑλένη Κλειώ χελιδών ἕλιξ μήτηρ Θέτις λαῖλαψ. (Τέχνη γραμματική *GG* I.I 15.1–16.1)

The order in which the neuter nouns are dealt with on the basis of their termination corresponds with the arrangement of the terminations of neuters in the Τέχνη γραμματική (*GG* I.I 16.2–3 οὐδετέρων δὲ ἕξ· α ι ν ρ σ υ, οἷον ἅρμα μέλι δένδρον ὕδωρ δέπας δόρυ ('And (the terminations) of neuter nouns are six: α ι ν ρ σ υ, for example ἅρμα μέλι δένδρον ὕδωρ δέπας δόρυ')), but with the difference that the termination -υ is dealt with third instead of last. This variation is probably due to the fact that Herodian wanted to deal with all the vowel terminations together at the beginning.[29]

---

25 The modern term 'nominal' is a cover term for both nouns and adjectives; it is used to translate the Greek term ὄνομα, which similarly covers both nouns and adjectives where no further specification is given.

26 See also the discussion in Egenolff (1887) 8.

27 For a treatment of the history of the parts of speech, see Wouters/Swiggers (2014).

28 The term ἄρθρα refers both to the article and the relative pronoun.

29 Cf. Egenolff (1887) 8.

Although the idea of the division of the terminations of nouns in the nominative singular according to the gender goes back to Protagoras,[30] the Τέχνη Γραμματική is probably the earliest known grammatical text to contain this distinction.[31] Finally, the similarities in the treatment of masculine, feminine, and neuter nouns on the basis of their terminations between Pseudo-Arcadius'-*Epitome* and the Τέχνη Γραμματική, suggest that Herodian was aware of, and followed (with the exception of minor variations) the categorization that we find in the Τέχνη Γραμματική. Herodian's two contributions to scholarship in this respect, were, on the one hand, that he did not consider necessary to treat separately the common masculine and feminine terminations, and on the other hand, that in the case of neuters he considered important to distinguish between the vowel and consonant terminations, and deal with all the vowel terminations together at the beginning, before the consonant terminations.

The use of the Περὶ καθολικῆς προσῳδίας in the Byzantine period

The Περὶ καθολικῆς προσῳδίας, though a work on accentuation, was employed by a large number of late antique and Byzantine scholars, not only those writing on accentuation but also on other aspects of scholarship, as for example on declension, orthography, and etymology.[32] Among these later authors and works are: Stephanus of Byzantium, Choeroboscus, Theognostus, the *Etymologicum genuinum*, the *Etymologicum Gudianum*, the *Etymologicum magnum*, and the *Epimerismi homerici.*[33] Herodian was consulted and cited not only as an authority in accentuation but also employed as a reference work for his large lists of examples arranged according to their termination. The encyclopaedic content of the Περὶ καθολικῆς προσῳδίας, with long lists of examples accompanied by their meanings and passages illustrating their use, despite encouraging the creation of epitomes and finally bringing to an end the transmission of the original work, made it attractive to later scholarship, which employed and perceived the material from various perspectives. The late antique and Byzantine works that derived material from Herodian have either incorporated or adjusted sections of Herodian's accentuation rules, or have transformed them into orthographical or declensional ones, in order to serve their purposes. Herodian is very often mentioned by name in these works, but even in cases where Herodian

---

30 See the discussion in Pfeiffer (1968) and Lallot (1998).

31 Lallot (1998) 26 argues in favour of the attribution of sections 6–10 (where the above-cited passage also belongs) of the Τέχνη Γραμματική to Dionysius Thrax.

32 Cf. Hunger (1967) 30. On the case of Theognostus specifically, see Galland (1882) 33.

33 See Lentz (1867) cxxxvi-cxcii, cciv-ccviii, ccxiii-ccxv; Dyck (1993) 777–78; and Roussou (2018) section 2.2.2. Xenis (2015b) has pointed out that Michael Syncellus is another witness to the ΠΚΠ, but our Epitome is not closely related to Syncellus' Περὶ τῆς τοῦ λόγου συντάξεως.

is not mentioned explicitly, the close verbal similarities and the use of common examples, sometimes even in the same order, suggest strongly that the material has been derived from Herodian. Although these late antique and Byzantine sources derive material from other grammarians as well, Herodian is the grammarian cited most frequently, and he is not mentioned in the same context or compared with other grammarians. To the extent that these later grammatical and lexicographical works derive material from Herodian, they can also be employed as indirect witnesses to Herodian's text, and they are indeed very often proved to be very useful for the reconstruction of the text of Pseudo-Arcadius' *Epitome*. The extensive employment of the Περὶ καθολικῆς προσῳδίας by later scholars illustrates the great impact and the significance of Herodian and his *magnum opus* in the history of scholarship.

New readings

We will now focus on some passages from Pseudo-Arcadius' *Epitome* where it is possible to propose new readings. We will focus particularly on personal names, city names, ethnic designations, and rare words and *hapax legomena*.

Personal names

Personal names are an important category of words employed by Herodian for his discussion of accentuation rules. The personal names that we find in the *Epitome* are either: (i) literary names or mythological names such as Πυριφλεγέθων (Homer), Χαρναβῶν (Sophocles fr. 604), Ἕρμων (Thucydides), Αἴγων (Theocritus), Φαέθων, Ἀπόλλων, Αἵμων; (ii) names of famous ancient authors such as Ὅμηρος, Ἀριστοφάνης, Θουκυδίδης; or (iii) names attested in inscriptions, such as Μύδιος (*LGPN* II)[34] and Δάτυννος (*LGPN* IIIB).

In passage 1 we have an example of a personal name coming from a literary source, namely Φώτυος. This passage contains a rule on words ending in –υος (with upsilon) that have more than two syllables and are proparoxytone:

> (1) Τὰ εἰς -ος ὑπερδισύλλαβα παραληγόμενα υ συνεσταλμένῳ προπαροξύνεται· σίκυος, Πέρυος, Φώτυος. τὸ δὲ Τιτυός ὀξύνεται καὶ τὸ υ οὐκ ἔχει συνεσταλμένον.
>
> (Words) ending in -ος that have more than two syllables and have a short υ in their penultimate syllable, are proparoxytone, e.g., σίκυος, Πέρυος, Φώτυος. But Τιτυός is oxytone and does not have a short Y. (Pseudo-Arcadius, *Epitome* 164.10 – 12)

Instead of Φώτυος Schmidt (46.21–23) printed Φώτιος with an iota, but the accentuation rule prescribes that there should be an υ in the penultimate syllable. We

34 This is the abbreviation for Fraser *et al.* (1987–).

get a hint about where this name came from, from the palimpsest fragment (cod. Vindob. Hist. gr. 10, f. 6v). The palimpsest contains the reading φωστυος, together with the information that this is the name of the leader of the Χάονες:

(1a) Φώστυος ἄρχων Χαόνων ὡς Θουκυδίδης ἐν β΄.

Φώστυος leader of the Χάονες as Thucydides in book two. (Palimpsest fragment, f. 6v)

The palimpsest is in general more extensive than Pseudo-Arcadius' *Epitome*, having more examples, some explanations (as in passage 1a), and select quotations from ancient authors. There is one place in literature where we are told about the leader of the Χάονες, Thucydides' *Hist.* 2.80.5:

(1b) καὶ αὐτῷ παρῆσαν Ἑλλήνων μὲν Ἀμπρακιῶται καὶ Ἀνακτόριοι καὶ Λευκάδιοι καὶ οὓς αὐτὸς ἔχων ἦλθε χίλιοι Πελοποννησίων, βάρβαροι δὲ Χάονες χίλιοι ἀβασίλευτοι, ὧν ἡγοῦντο ἐπετησίῳ προστατείᾳ ἐκ τοῦ ἀρχικοῦ γένους Φώτυος[1] καὶ Νικάνωρ.

And Ambraciots, Anactorians, Leucadians among the Hellenic troops stood by him, and one thousand Peloponnesians with whom he arrived, and one thousand Chaonian barbarians, who were not ruled by a king but were led by Photyos and Nicanor, (members) of the ruling family, who were in charge of the yearly leadership. (Thucydides, *Hist.* 2.80.5, Alberti)[35]

Herodian must have derived the name from Thucydides. The manuscripts of Thucydides (at *Hist.* 2.80.5) transmit the name Φώτυος, who is said to be the leader of the Χάονες. Interestingly, there is also a textual problem in Thucydides. Two of Thucydides' manuscripts have the correction Φώτιος (one of them by the first hand and the other by a third hand). This change was prone to happen because Φώτιος is a well-known Greek name, while Φώτυος, which was a barbarian name, was not common. Most editions of Thucydides, including the Oxford Classical Texts edition, print Φώτιος, but Alberti more recently (passage 1b) printed Φώτυος. The reading Φώτυος in Pseudo-Arcadius' *Epitome* and the palimpsest reading Φωστυος, which must be a corrupt form of Φώτυος, in the context of a rule ending in -υος speak in favour of reading Φώτυος in Thucydides. At the same time the reading Φώτυος in Pseudo-Arcadius' *Epitome* (as well as the palimpsest reading Φωστυος) show that Thucydides, among other authors, was a source from which Herodian derived material.

Another example of a literary name comes from passage 2, namely Ὀρφήν, a Doric alternative form for Ὀρφεύς, which is a rare name attested only in Ibycus.

(2) Τὰ εἰς -ην λήγοντα μὴ συντεθειμένα ἀπὸ τῶν εἰς ΗΝ ῥητῶν, εἰ ἔχοι πρὸ τοῦ -ην δασὺ σύμφωνον ἢ ψιλόν, ὀξύνεσθαι θέλει, οἷον· Ἀρχήν, αὐχήν, †ναυχήν, κηφήν, Ὀρφήν.

---

1 ῥητῶν Schmidt: ῥημάτων MO   ἔχοι M: ἔχει O   2 Ἀρχήν MO; ἀχήν Lobeck   αὐχήν ναυχήν MO; λειχήν αὐχήν Lobeck; fortasse αὐχήν παρὰ τὸ αὐχμήν coll. *Et. Gen.* α 1432.1 vel αὐχήν παρὰ τὴν αὐχμήν coll. *EM* 174.21

---

35 Alberti (1972–2000).

(Words) ending in -ην that are not compounded from (words) which can be said individually, if they have before the -ην an aspirated consonant or an unaspirated, are usually oxytone, e.g., Ἀρχήν, αὐχήν, †ναυχήν, κηφήν, Ὀρφήν (Ibyc. fr. 306) (Pseudo-Arcadius, *Epitome* 115.11–13).

(2a) ὀνομάκλυτον Ὀρφήν.

'Orphen of famous name' (Ibyc. fr. 306)

Ἀρχήν in passage (2) is also a personal name, but, as far as we know, it is not a literary one. Rather, it is attested in epigraphic sources (*LGPN* IIIA). Schmidt (6.10–13) did not realise that Ἀρχήν and Ὀρφήν are proper names, and did not capitalise them. Since the name Ὀρφήν is attested only once in literature, namely in Ibycus, it is very likely that Herodian derived this name from him, although we cannot exclude the possibility that this name was attested in other literary sources that have not come down to us. This further suggests that Herodian most probably derived material from lyric poetry among his other literary sources.

In passage 3, following the rule on disyllabic common nouns ending in -αυλος which are oxytone, we are given some personal names. Among the cited personal names, the majority of Pseudo-Arcadius' manuscripts transmit the name Σαῦλος:

(3) Τὰ εἰς -αυλος δισύλλαβα μονογενῆ μὴ κύρια ὀξύνεται· αὐλός, δαυλός, καυλός. τὸ δὲ Βραῦλος, Παῦλος, Σαῦλος κύρια. ⟨....⟩ καὶ τὸ ναῦλος μὴ ὂν κύριον. τὸ δὲ φαῦλος ἐπίθετον.

2 Σαῦλος M: σαύλος O   ⟨...⟩] προπερισπᾶται add. Schmidt (app. crit.)

Disyllabic (words) ending in -αυλος that have one form for all genders and are not proper names, are oxytone, e.g., αὐλός, δαυλός, καυλός. But Βραῦλος, Παῦλος, Σαῦλος are proper names. ⟨...⟩ and ναῦλος although it is not a proper name. And φαῦλος is an adjective. (Pseudo-Arcadius, *Epitome* 180.1–3)

Instead of Σαῦλος, Schmidt (59.18–21) printed the reading of manuscript A, Δαῦλος, but there is not enough evidence for Δαῦλος as a personal name. Although Σαῦλος is a personal name which usually occurs in religious texts, I have kept it in the text of Pseudo-Arcadius' *Epitome*. The *Epitome* has a small number of religious words, which are probably post-Herodianic, although we cannot be sure if these words were added to Herodian's text before or after the *Epitome* was created. Given the great size of the Περὶ καθολικῆς προσῳδίας and the large number of examples contained therein, it is generally assumed that the author of the *Epitome* would not have added examples of his own to exemplify Herodian's accentuation rules but would have selected a small number of examples from Herodian's large lists of examples. If we thus accept that the religious words in the *Epitome* were not added by its composer, we will have to attribute them to the textual transmission, namely to the scribes copying Herodian's text or the *Epitome*. The religious words in Pseudo-Arcadius' *Epitome* suggest that Herodian's text or the *Epitome*'s text was copied and transmitted by Christian scribes or scribes with theological knowledge and interests or

Greek-speaking Jews. The religious words that we find are for example ὀσσήν in passage 3a (whose stem recalls the Osseni, a Jewish sect) and θεοτόκος in passage 3b:

> (3a) Τὰ εἰς -ην ἁπλᾶ ἔχοντα δεδιπλασιασμένον σύμφωνον ὀξύνεται· ἐσσήν, ὀσσήν, Τελλήν, βαλλήν, πλὴν τοῦ Ἕλλην.
>
> Uncompounded (words) ending in -ην that have a double consonant are oxytone: ἐσσήν, ὀσσήν, Τελλήν, βαλλήν, except for Ἕλλην. (Pseudo-Arcadius, *Epitome* 116.7–8)

> (3b) Τὰ ἐπὶ ἐνεργείας τιθέμενα συντεθειμένα ἀπὸ τοῦ 'τέμνω, φονεύω, ὀρύσσω, κτείνω, νέμω, τίκτω' παροξύνονται· σκυτοτόμος, μιαιφόνος, τοιχωρύχος, πατροκτόνος, οἰκονόμος, θεοτόκος.
>
> Compound (words) that are used of (someone undertaking) an action derived from the (verbs) τέμνω φονεύω ὀρύσσω κτείνω νέμω τίκτω are paroxytone: σκυτοτόμος, μιαιφόνος, τοιχωρύχος, πατροκτόνος, οἰκονόμος, θεοτόκος. (Pseudo-Arcadius, *Epitome* 228.1–3)

The Osseni (Ὀσσηνοί) are otherwise attested in the ecclesiastical author Epiphanius (e.g., in his *Panarion* I 222.14, I 336.2) in the 4th century, while the word θεοτόκος is attested for the first time in the Septuagint (Ode 9). It would be odd for Herodian to have cited words with religious content (especially Christian and Jewish). Herodian was part of the emperor's court and dedicated his Περὶ καθολικῆς προσῳδίας to him. It is reasonable to assume that Herodian would have avoided citing religious words that pointed to Christianity, in order not to enrage the emperor Marcus Aurelius who persecuted the Christians.[36] At the same time, the Jewish-Roman tensions which resulted in several Jewish-Roman wars in the first and second centuries CE would have been another reason for Herodian to avoid citing words related to Judaism, so as not to disgruntle the emperor. The juxtaposition of Παῦλος and Σαῦλος suggests that the person meant was the follower of Jesus, Saul, afterwards known as Paul (Acts 13:9), and this further suggests that Παῦλος Σαῦλος must have come from someone familiar with the New Testament. Pseudo-Arcadius' *Epitome* is thus a text with material from different chronological layers.

In passage 4, in the rule on recessive disyllabic masculines ending in -θων, among the cited examples we find Μάθων:

> (4) Τὰ εἰς -θων δισύλλαβα ἀρσενικὰ βαρύνεται, ὁπότε μὴ ἐθνικὰ εἴη ἢ περιεκτικά· μόθων, ὤθων (?),[37] † Μάθων (ὁ Θηβαῖος), † πίθων (ὁ πίθηκος ὑποκοριστικῶς), Ῥίνθων, αἴθων. τὸ δὲ Πυθών πόλις καὶ Σιθών καὶ πιθών καὶ κιθών (ὁ χιτών) ὀξύνεται.
>
> ———————
>
> 1 ἢ περιεκτικά M: om. O   μόθων Schmidt: μώθων M: μύθων O   ὤθων MO; κώθων Lobeck; ῥώθων Schmidt (Corrigenda)   2 Μάθων – Θηβαῖος] possis μάλθων (ὁ θηλαῖος/ θήλαιος) coll.

---

36 On the persecutions against the Christians during the time of Marcus Aurelius, see Edwards (2012).

37 This is nowhere else attested. Since this reading is doubtful, I have noted a question-mark in brackets.

Theodos. Περὶ κλίσεως τῶν εἰς ΩΝ βαρυτόνων 19.24–29 et *Su.* μ 111 πίθων – πίθηκος Lobeck: πήθων ὁ πήθικος (πηθικός M[a.c.]) MO 3 πιθών M: τιθών O

Disyllabic masculines ending in -θων are recessive, when they are not ethnic designations and they do not denote a place in which things are situated, e.g., μόθων, Ὄθων, Μάθων ('Theban'), πίθων ('ape', diminutively), Ῥίνθων, αἴθων. But Πυθών (a city) and Σιθών and πιθών and κιθών ('tunic') are oxytone. (Pseudo-Arcadius, *Epitome* 121.1–4)

If indeed Μάθων was the intended word here one would expect that this was a famous enough person to be mentioned by Herodian. But no Theban figure with this name is well known, and this name is not attested in a literary source. The *Lexicon of Greek Personal Names* has no-one of this name.[38] Moreover, even if there was such a personal name and there was a person with this name who was famous enough to be cited, still no other name in this *Epitome* is accompanied by an explanatory gloss like this; only common nouns and adjectives are given such glosses. If this is indeed a personal name it must be someone obscure whom we do not otherwise know, or it might have been a known figure attested in a literary source that has not come down to us. A parallel passage from Theodosius, passage 4a, suggests the emendation of Μάθων into μάλθων:

(4a) Τὰ εἰς -ων λήγοντα δισύλλαβα ἔχοντα πρὸ τοῦ ω ἕν τι τῶν δασέων διὰ τοῦ ω κλίνεται, οἷον μόθων μόθωνος (σημαίνει τὸν πανοῦργον), κάνθων κάνθωνος, ⟨μάλθων μάλθωνος⟩ (ὁ μαλακός), Ῥίνθων Ῥίνθωνος (ὄνομα κύριον), σίφων σίφωνος, κύφων κύφωνος, Τύχων Τύχωνος (ὅ ἐστι δαίμων ὁ περὶ τὴν Ἀφροδίτην), Βρύχων Βρύχωνος, πίθων πίθωνος (οὕτως ὁ πίθηκος ὑποκοριστικῶς)· καὶ τὸ γλήχων γλήχωνος βαρυνόμενον φυλάττει τὸ ω, εἰ καὶ παρ' Ἀττικοῖς ὀξύνεται καὶ θηλυκῶς λέγεται.

Disyllabic (words) ending in -ων that have one of the aspirated consonants before the ω, e.g., μόθων μόθωνος (it means 'cunning'), κάνθων κάνθωνος, ⟨μάλθων μάλθωνος⟩ ('soft'), Ῥίνθων Ῥίνθωνος (a proper name), σίφων σίφωνος, κύφων κύφωνος, Τύχων Τύχωνος (who is a god, the one connected to Aphrodite), Βρύχων Βρύχωνος, πίθων πίθωνος (thus the ape diminutively); and when γλήχων γλήχωνος is recessive it maintains the omega, even if in the Attic it is oxytone and feminine. (Theodosius, Περὶ κλίσεως τῶν εἰς -ων βαρυτόνων 19.24–29)

In passage 4a μάλθων μάλθωνος has been supplied by the editor on the basis of the transmitted gloss ὁ μαλακός and on the basis of the fact that this passage discusses words ending in -θων, -φων, and -χων. In passage 4, the corruption of μάλθων to μάθων must have been easy as it involves the omission of a single letter. This emendation, however, would not correspond to the word's transmitted gloss, Θηβαῖος, but Θηβαῖος might also be the corrupt form of another gloss. Passage 4a from Theodosius provides us with the gloss μαλακός (meaning 'soft, gentle, mild') for μάλθων ('weakling, weak person or animal'). The *Suda* (passages 4b and 4c) twice provides us with the quotation ὁ γύνανδρός τε καὶ μάλθων τύραννος "the womanish and weakling tyrant":

---

38 Fraser *et al.* (1987-) volume IIIB.

(4b) Μάλθων· 'ὁ γύνανδρός τε καὶ μάλθων τύραννος'. ἀντὶ τοῦ ἔκλυτος.

Μάλθων ('weakling'): 'the womanish and weakling tyrant'. For 'mild'. (*Suda* M 111)

(4c) Γύνανδρος: κίναιδος. 'ὁ γύνανδρός τε καὶ μάλθων τύραννος'. ἀντὶ τοῦ ἔκλυτος.

Γύνανδρος ('womanish'): catamite. 'The womanish and weakling tyrant'. For 'mild'. (*Suda* Γ 494)

The juxtaposition of γύνανδρος and μάλθων suggests that the words have a similar meaning. Nevertheless, neither μαλακός nor γύνανδρος are palaeographically close to the transmitted Θηβαῖος, and so it is not easy to explain the corruption of either μαλακός or γύνανδρος into Θηβαῖος. Perhaps the gloss for μάλθων was θηλαῖος, a relative of θῆλυς, occurring in Pseudo-Arcadius' *Epitome* (167.1), that could have been corrupted into Θηβαῖος by change of just one letter. θηλαῖος is rare and is attested only in Byzantine Greek,[39] but this might be an early occurrence of the word.

City names

Another category of words that Herodian cites as examples for his accentuation rules are city names, as for example: Σιδών, Βαβυλών, and Αὐλίς. The city names that are cited in Pseudo-Arcadius' *Epitome* are sometimes attested in geographical sources (e.g., Σιδών at Strabo 1.2.33 (p. 40C.36), Βαβυλών at Strabo 1.1.12 (p. 7C.19), Αὐλίς at Strabo 7.3.6 (p. 298C.20)),[40] but sometimes in literature as well, as for example Ὀλοσσών which is attested in Homer (*Il.* 2.739). The epitome contains also city names from Italy (e.g., Σούλμων at 16.14–15), from Lydia (e.g., Τράλλις at 32.13), from Libya (e.g., Χάλκη at 122.12), from Ionia (e.g., Κλαζομεναί at 127.16–17), and other areas outside Greece.

In passage 5, following the citation of recessive words ending in –υρα with more than two syllables, we are given a city name which is oxytone:

(5) Τὰ εἰς -ρα ὑπὲρ δύο συλλαβὰς τῷ υ παραληγόμενα βαρύνεται. ἀλλ' εἰ μὲν ἔχωσι τὸ Α μακρόν, παροξύνεται· πορφύρα, Ἐφύρα· εἰ δὲ βραχύ, προπαροξύνεται· Κέρκυρα, ἄγκυρα, γέφυρα. τὸ δὲ Γλαφυραί ὀξύνεται, εἴτε ὄνομα πόλεως εἴη, εἴτε ἐπιθετικόν.

(Words) ending in -ρα that have more than two syllables and have υ in their penultimate syllable, are recessive. But if they have a long α, they are paroxytone, e.g., πορφύρα, Ἐφύρα; but if it is short, it is proparoxytone: Κέρκυρα, ἄγκυρα, γέφυρα. But Γλαφυραί is oxytone, whether it is the name of a city or an adjective. (Pseudo-Arcadius, *Epitome* 242.9–12)

39 See Trapp *et al.* (1994–) s.v. θήλαιος. The accent of the word is different, but it is possible that both accentuations were in use.

40 I cite Strabo from the edition of Radt (2002–2005).

All manuscripts of Pseudo-Arcadius' *Epitome* transmit the form Γλαφυραί, with the exception of C which has the singular form. The city name in question is used in the plural, and so it makes sense for it to be cited by Herodian in its usual, plural form rather than in the singular. The citation of the city name in the plural does not create any problems for making the point that there is also the adjective γλαφυρά, the feminine of the adjective γλαφυρός ('polished, exact'), which is a homonym. Schmidt (115.12–17) printed the reading of the least reliable manuscript, which presents this example in the singular. Γλαφυραί is a *lectio difficilior*, and it seems that the scribe of C intervened in changing the plural into singular, thinking that this would make the correspondence with the adjective more clear and straightforward. This city name is attested in both literature and geographical sources, and either could have been the source for Herodian. West, in his edition of Homer (*Il.* 2.711–12) accents this city name on its penultimate syllable. However, Herodian might have had a different view on the accentuation of this word, or the text of Homer which Herodian consulted might have had this city name as oxytone.

In passage 6, Herodian outlines the rule on words ending in -τη that have more than two syllables and have two consonants which belong to the same syllable. Manuscripts M and O provide us with the new reading Ἀκέστη, which has been omitted by the rest of the manuscript tradition:

(6) Τὰ εἰς -τη ὑπερδισύλλαβα ἔχοντα κατὰ σύλληψιν σύμφωνον, ἢ πρὸ τοῦ τέλους τὸ ρ, κύρια ὄντα βαρύνεται· Ἀδρήστη, Ἰοκάστη, Ἀκέστη.

---

1 τοῦ O: om. M 2 Ἀδρήστη M: ἀδράστη O Ἰοκάστη Schmidt: ἰωκάστη MO

(Words) with more than two syllables that end in -τη and have a consonant in conjunction, or (the letter) P in their penultimate syllable, when they are proper names, they are recessive, e.g., Ἀδρήστη, Ἰοκάστη, Ἀκέστη. (Pseudo-Arcadius, *Epitome* 260.13–14)

The city name Ἀκέστη occurs later in Hesychius (O 2055) and in Stephanus of Byzantium (*Suda* A 159). Ἀκέστη is not attested anywhere else before Herodian, and so it is a puzzle where this name came from, although a geographical source seems to be the most likely source.

Ethnic designations

Herodian includes ethnic designations among the examples for his accentuation rules. Apart from the very common ethnic designations, as for example Ἕλλην, Ἀθηναῖος, Θηβαῖος, and Ῥωμαῖος, we find some which are rare. For example, passage 7 contains an ethnic designation which is otherwise attested only in Stephanus of Byzantium.

In passage 7, in a rule on words ending in -αν that are not compounds, the manuscripts transmit the readings δωρεάν and δηρεάν. The reading δηρεάν is a non-existing word, while δωρεάν ('freely') is an adverb (or it could otherwise be considered to be the accusative of the noun δωρεά ('gift')) and is unsuitable in the context of this rule which deals with the nominative of nominals ending in -αν. Lobeck proposed two emendations: (i) Δαρειάν, an alternative form for Δαρεῖος, the Persian King, and (ii) Δωριάν, an alternative for Δωριεύς ('Dorian'), a rare ethnic designation attested in Stephanus of Byzantium (*Suda* Δ 149).[41] Δωριάν is closer to the transmitted reading, and is to be preferred:

(7) Τὰ εἰς -αν λήγοντα ἁπλᾶ ὀνόματα ὀξύνεσθαι θέλει, ἐπ' εὐθείας πτώσεως μὴ τροπὴν ἀναδεξάμενα ἀπὸ τῶν εἰς -ην βαρυτόνων, οἷον· παιάν, Ἰάν, Δωριάν, Ἀλκμάν, τιτάν, πελεκάν.

1 μὴ M: μήτε O   2 Δωριάν Lobeck: δηρεάν vel δωρεάν M: δωρεάν O; Δαρειάν Lobeck

Uncompounded nominals ending in -αν are usually oxytone, in the nominative case if they have not been modified from recessives ending in -ην, e.g., παιάν, Ἰάν, Δωριάν, Ἀλκμάν, τιτάν, πελεκάν. (Pseudo-Arcadius, *Epitome* 115.4–6)

For the formation of Δωριάν one may compare the ethnic designation Αἰνιάν (Stephanus of Byzantium A 134), an alternative for Αἴνιος, designating a person from Αἰνία.[42]

Passage (8) comes from book one of Pseudo-Arcadius' *Epitome* and it contains a rule on words with more than two syllables which end in -δων and are oxytone. Among the cited examples we have the cities Ἀμυδών and Καλυδών, and the ethnic designation Ἀπιδών. The final example in the text as transmitted is χελιδών, which is not a city name. One may thus assume that this reading was added to the text at a later stage by mistake or that there is a lacuna in the text.

(8) Τὰ εἰς -δων ὑπερδισύλλαβα ὀξύνεται, παραλήγοντα τῷ υ ἢ ι, ὀνόματα ὄντα πόλεων ἢ ἐθνικά· Ἀμυδών, Καλυδών, Ἀπιδών, ⟨...⟩ χελιδών.

1 Υ ἢ Ι Schmidt: ι ἢ MO   ἢ ἐθνικά $M^{p.c.}$O: om. $M^{a.c.}$ 2 Ἀπιδών Lehrs: ἀσπιδών MO

(Words) ending in -δων that have more than two syllables are oxytone when they have υ or ι in their penultimate, and they are names of cities or ethnic designations: Ἀμυδών, Καλυδών, Ἀπιδών, ⟨...⟩ χελιδών. (Pseudo-Arcadius, *Epitome* 120.11–13)

According to a scholion to Aristophanes' *Birds* 1680,[43] Herodian mentioned in book one of his Περὶ καθολικῆς προσῳδίας that barbarians are called χελιδόνες in the masculine:

(8a) οὕτω δὲ αὐτό φησι βαρβάρως καὶ δυσφράστως, ὥσπερ αἱ χελιδόνες. καὶ Αἰσχύλος τὸ βαρβαρίζειν χελιδονίζειν φησί (Cf. A. *Ag.* 1050 and fr. 450 Radt), καὶ Ἴων ἐν Ὀμφάλῃ (*TrGF* 19 F 33

41 Lobeck (1837) 190 n.2.

42 On Αἰνιάν, see also Fraser (2009) 95, 132. Cf. Ἀγριᾶνες, a variant for the ethnic Ἀγρίαι (Stephanus of Byzantium α 47).

43 Holwerda (1991).

Snell) τοὺς βαρβάρους, χελιδόνας ἀρσενικῶς φησιν, ὡς Ἡρωδιανὸς ἐν τῷ πρώτῳ τῆς καθόλου φησίν.

Thus he says this barbarously and with poor expression, just like swallows. And Aeschylus calls 'speaking a foreign language' χελιδονίζειν ('twittering like a swallow'), and Ion in the *Omphale* calls barbarians χελιδόνες in the masculine, as Herodian says in the first (book) of his Περὶ καθολικῆς προσῳδίας. (sch. Ar. *Av.* 1680)

Thus, although the occurrence of χελιδών in the context of passage 8 creates the suspicion that χελιδών is out of place, the evidence from the Aristophanic scholion for the citation of χελιδών in book one of the Περὶ καθολικῆς προσῳδίας encourages us to think that χελιδών belongs to the passage in question. Passage 8a makes mention of Aeschylus and refers to a point in the *Agamemnon* (passage 8b):

(8b) **Χο.** σοί τε λέγουσα παύεται σαφῆ λόγον·
ἐντὸς δ' ἁλοῦσα μορσίμων ἀγρευμάτων
πείθοι' ἄν, εἰ πείθοι'· ἀπειθοίης δ' ἴσως.
Κλ. ἀλλ' εἴπερ ἐστὶ μὴ χελιδόνος δίκην
ἀγνῶτα φωνὴν βάρβαρον κεκτημένη,
ἔσω φρενῶν λέγουσα πείθω νιν λόγωι.

Chor. It is to thee she has spoken, and now she has finished, a plain speech.
Taken as thou art within the toils of fate,
pray obey her, if thou wilt obey; though perhaps thou wilt disobey.
Cl. Nay, if she is not, like a swallow,
possessed of an unintelligible foreign tongue,
speaking ... I attempt (?) to persuade her by my words. (Aeschylus. *Ag.* 1047–1052)[44]

At this point of the play Cassandra remains silent and Clytemnestra is wondering if Cassandra is a foreigner and if she does not understand Greek. Clytemnestra uses the phrase χελιδόνος δίκην. The unintelligible foreign language is compared with the twittering of a swallow.[45] The Aristophanic scholion (passage 8a) suggests that a phrase similar to χελιδὼν δὲ ὁ βάρβαρος ἀρσενικῶς might have originally been present in passage 8 or at least in the original Περὶ καθολικῆς προσῳδίας. Herodian appears to have derived material from a learned tradition, in which Aeschylus was interpreted in such a way that χελιδών meant βάρβαρος. Thus, χελιδών could be taken as a sort of ethnic designation, and this could explain its presence in the context of the accentuation rule in passage 8.

44 Translated by Fraenkel (1950).

45 That the twittering of the swallow was used of barbarous tongues by the Greeks can be also seen in Hesychius (X 325) and in the *Suda* (X 187).

Rare words and hapaxes

In Herodian's text we read a number of rare words and *hapax legomena*, that is to say words that are nowhere else attested in the texts that are known to us today. Some words transmitted in the manuscripts of the *Epitome* that are nowhere else attested sometimes raise the question whether they should be regarded as *hapax legomena* or scribal errors, and it is often very difficult to draw the line between the two possibilities. In the case of rare words, Pseudo-Arcadius' *Epitome* is often the earliest attestation and we cannot know where Herodian derived them from, although one may assume that he must have used earlier lexicographical works.

Passage 9 contains a rule on the recessive comparatives and hypocorisms ending in -ων. The manuscripts M and O provide us with the new good reading μωρίων, which allows us to replace the non-existing word μωτίων, which was printed by the two previous editions of Pseudo-Arcadius' *Epitome*.

(9) Τὰ εἰς -ων συγκριτικὰ καὶ ὑποκοριστικὰ βαρύνεται· ⟨...⟩ Ἀτρείων, Κρονίων, μωρίων, Ἡφαιστίων.

———————

1 συγκριτικὰ – ὑποκοριστικὰ] debemus addere καὶ πατρωνυμικὰ vel scribere πατρωνυμικὰ pro συγκριτικὰ ὑποκοριστικὰ M: ὑποκορυστικὰ O βαρύνεται M: βαρύνονται O

Comparatives and diminutives ending in -ων are recessive: ⟨...⟩ Ἀτρείων, Κρονίων, μωρίων, Ἡφαιστίων. (Pseudo-Arcadius, *Epitome* 129.1–2)

μωρίων is a rare word meaning 'fool', attested again in the scholia to Dionysius Thrax (scholia Vaticana *GG* I.III 227.8–9, scholia Marciana *GG* I.III 376.8, scholia Londinensia *GG* I.III 539.6).[46] The later attestations show that this word was not made up and that this is an existing word, although we cannot know what Herodian's source was.

In passage 10 we have a rule on recessive words which have more than one syllable and end in -ξ. This rule contains the *hapaxes* βλεφαροσπάξ and λιτροβαστάξ:

(10) Πᾶν ὄνομα εἰς ξ λῆγον ὑπὲρ μίαν συλλαβὴν βαρύνεται, χωρὶς εἰ μὴ παρὰ ῥῆμα εἴη συντεθειμένον καὶ μίαν συλλαβὴν φυλάττοι τοῦ ῥήματος, καὶ ἐπιθετικὸν ὑπάρχοι, μὴ ἐθνικόν· Φαίαξ, θώραξ, ἄναξ, κλῖμαξ, αὖλαξ· τὸ διασφάξ ⟨ἀπὸ ῥήματος⟩ ἢ ἀπὸ τοῦ σφάξ ὀξυτόνου. τὰ δὲ ἀπὸ ῥήματος συγκείμενα· βλεφαροσπάξ καὶ λιτροβαστάξ καὶ νεκροβαστάξ. τὸ δὲ Ἄσφαξ βαρύνεται ὡς ἐθνικόν.

———————

2 φυλάττοι M: φυλάττει O ὑπάρχοι M: ὑπάρχει O 3 κλῖμαξ αὖλαξ Schmidt: κλίμαξ, αὔλαξ MO ἀπὸ ῥήματος addidi coll. *Ep. Hom.* 7 B et *Et. Gud.* 132.5–7 Stef. 4 βλεφαροσπάξ Hermann: βλεφαροπάξ M: βλεφυροπάξ O; βλεφαροσπάξ vel βλεφαροστάξ Lobeck λιτροβαστάξ MO; λητροβαστάξ Schmidt, ut Bloch de A notavit; φιτροβάσταξ Lobeck

———

46 The scholia to Dionysius Thrax have been edited by Hilgard (1901).

Each nominal ending in ξ that has more than one syllable is recessive, unless it is compounded from a verb and it maintains one syllable from the verb, and it is an adjective, but not an ethnic designation: Φαίαξ, θώραξ, ἄναξ, κλῖμαξ, αὖλαξ; διασφάξ ⟨from a verb⟩ or from the oxytone σφάξ. And those that are composed from a verb: βλεφαροσπάξ and λιτροβαστάξ and νεκροβαστάξ. But Ἄσφαξ is recessive as it is an ethnic designation. (Pseudo-Arcadius, *Epitome* 130.7–12)

βλεφαροσπάξ is an emendation of the reading βλεφαροπάξ transmitted in manuscripts M and A. Manuscript O has βλεφυροπάξ. Although βλεφαροσπάξ and λιτροβαστάξ are nowhere else attested, they must be real words. By analysing βλεφαροσπάξ in its two members, βλέφαρον and σπάω, one can make sense of the word meaning 'arching the eyebrows'. Schmidt (18.16–22) printed λητροβαστάξ with an η, thinking that this was the reading of manuscript A. The collation of A which Schmidt used contained this mistake. One could make sense of λιτροβαστάξ by analysing it in its two members, λίτρον or λίτρα and βαστάζω, and could mean the person carrying, holding, or taking away (stealing) the λίτρον ('measure of capacity'; see LSJ[47] s.v.) or the λίτρα (1. 'a silver coin of Sicily', 2. 'a pound' (weight), 3. 'measure of capacity'; see LSJ s.v.).

Passage 11 contains a rule on compound words ending in -ρος that have more than two syllables and have ο in their penultimate syllable. All manuscripts of Pseudo-Arcadius' *Epitome* except for O transmit κατάμορος, while O has κατάκορος. But κατάμορος is not attested as such, but in its epic form, κάμμορος. Given this, it is unlikely that Herodian would have cited it in another form, which was not the usual one. Thus, I print κατάκορος instead of κατάμορος.

(11) Τὰ εἰς -ρος ὑπερδισύλλαβα παραληγόμενα τῷ ο σύνθετα ὄντα προπαροξύνεται· κατήγορος, συνήγορος, περὶ ὧν ἐν τῷ τέλει λέξομεν· νῦν δὲ διαληψόμεθα περὶ τῶν ἐχόντων τὸν τόνον ἐν τῇ προπαραληγούσῃ· ἄνορος, πάπορος, κατάκορος.

3 ἄνορος] ἄγορος vel ἄκορος Lobeck πάπορος] Πάκορος Lobeck; παντοπόρος Schmidt (app. crit.) κατάκορος O: κατάμορος M; κάμμορος vel κάμορος Lobeck; γαμόρος Schmidt (app. crit.)

Compound (words) ending in -ρος that have more than two syllables and have ο in their penultimate syllable are proparoxytone: κατήγορος, συνήγορος, about which we will talk in the end; now we will deal with those that have their accent in the antepenultimate: ἄνορος, πάπορος, κατάκορος. (Pseudo-Arcadius, *Epitome* 205.5–206.2)

Conclusions

In conclusion, we have seen that the text of Pseudo-Arcadius' *Epitome* can be reconstructed on the basis of good readings found in the manuscripts that are used for the first time in my edition and on the basis of parallel passages. Herodian represents primarily the systematization of the treatment of accentuation in the second century

47 This is the abbreviation for Liddell *et al.* (1996).

AD. Herodian derived material as examples for his accentuation rules from literary sources, including Thucydides, lyric poetry, Homer, and tragic poetry, and also probably from geographical and lexicographical sources. Herodian's use of literary examples for his rules often reinforces our knowledge of the textual transmission of these literary texts, and our ability to reconstruct them. Herodian's remarks on accentuation found their way into the Homeric, Aristophanic, and tragic scholia, through which we get a view of Herodian's lost works, as for example the Περὶ καθολικῆς προσῳδίας, the Περὶ Ἰλιακῆς προσῳδίας, the Περὶ Ὀδυσσειακῆς προσῳδίας, and the Περὶ παθῶν. Some post-Herodianic words, mostly religious ones found their way into Herodian's text. Pseudo-Arcadius' *Epitome* is a text with material from different chronological layers, including material from the New Testament. We are not sure if Herodian collected his examples directly from literary sources and from earlier geographical and lexicographical sources. Herodian's contribution must have been the organisation of material according to the termination of the words, since the termination of a word is crucial for its accentuation. In fact, the organisation of the material according to the termination made the material reusable for other aspects of grammar further than accentuation. Although we do not have much information about Herodian's predecessors in the field and his sources, the fact that Herodian is the grammarian mostly cited in a large number of late antique and Byzantine sources and the close verbal similarities between Herodian's text and these sources, illustrate Herodian's great impact on ancient and Byzantine scholarship. Herodian's Περὶ καθολικῆς προσῳδίας, although a work aiming mainly to prescribe accentuation, contained material that was useful to later scholarship, further than the constraints of accentuation. The authority of Herodian, being employed as reference point by grammatical and lexicographical sources for hundreds of years later, highlights his significance in the history of scholarship.[48]

48 I am deeply grateful to Philomen Probert and Eleanor Dickey for their feedback on my paper.

Eleanor Dickey

What does a Linguistic Expert Know? The Conflict between Analogy and Atticism

The phrase 'linguistic expert' ought to be an oxymoron. Language consists of a set of arbitrary conventions shared by members of a given community: it is simply whatever the members of that community do when they open their mouths, and therefore no member of the community should be able to have any special status in relation to it. Yet expertise is precisely a special status, one that arises from a community's belief that an individual has worthwhile knowledge not shared by most other people: an expert is not simply someone who knows something unusual, but someone whose unusual knowledge is recognised by others. How does a community decide that some people know more than others about something that consists only of what everyone does?

There are, of course, many answers to this question, for there are many different possible bases on which linguistic expertise can be claimed, depending on the community involved. Some cultures do not have the concept of linguistic expertise or linguistic correctness at all, and these cultures are not necessarily primitive or unsophisticated: it has been persuasively argued that Aeschylus's Athens was one of them.[1] Yet later Greek speakers clearly did have linguistic experts, and moreover the basis of their expert status seems to have changed over time. How did this situation arise?

Linguistic expertise based on knowledge of analogical rules

The earliest substantial evidence for the existence of linguistic experts in Greece can be seen in passage 1, from Aristophanes' *Clouds*, in which Socrates is presented as such an expert:

1)	So:	ἀλλ' ἕτερα δεῖ σε πρότερα τούτου μανθάνειν, τῶν τετραπόδων ἅττ' ἐστὶν ὀρθῶς ἄρρενα.
	Streps:	ἀλλ' οἶδ' ἔγωγε τἄρρεν', εἰ μὴ μαίνομαι· κριός, τράγος, ταῦρος, κύων, ἀλεκτρυών.
	So:	ὁρᾷς ἃ πάσχεις; τήν τε θήλειαν καλεῖς ἀλεκτρυόνα κατὰ ταὐτὸ καὶ τὸν ἄρρενα.

1 West (1990) 10 – 12.

Eleanor Dickey, University of Reading

https://doi.org/10.1515/9783110660982-008

Streps: πῶς δή, φέρε;
So: πῶς; ἀλεκτρυὼν κἀλεκτρυών.
Streps: νὴ τὸν Ποσειδῶ. νῦν δὲ πῶς με χρὴ καλεῖν;
So: ἀλεκτρύαιναν, τὸν δ' ἕτερον ἀλέκτορα.
Streps: ἀλεκτρύαιναν; εὖ γε, νὴ τὸν Ἀέρα·
ὥστ' ἀντὶ τούτου τοῦ διδάγματος μόνου
διαλφιτώσω σου κύκλῳ τὴν κάρδοπον.
So: ἰδοὺ μάλ' αὖθις, τοῦθ' ἕτερον· τὴν κάρδοπον
ἄρρενα καλεῖς θήλειαν οὖσαν.
Streps: τῷ τρόπῳ;
ἄρρενα καλῶ 'γὼ κάρδοπον;
So: μάλιστά γε,
ὥσπερ γε καὶ Κλεώνυμον.
Streps: πῶς δή; φράσον.
So: ταὐτὸν δύναταί σοι κάρδοπος Κλεωνύμῳ.
Streps: ἀλλ', ὦγάθ', οὐδ' ἦν κάρδοπος Κλεωνύμῳ,
ἀλλ' ἐν θυείᾳ στρογγύλῃ γ' ἀνεμάττετο.
ἀτὰρ τὸ λοιπὸν πῶς με χρὴ καλεῖν;
So: ὅπως;
τὴν καρδόπην, ὥσπερ καλεῖς τὴν Σωστράτην.
Streps: τὴν καρδόπην, θήλειαν;
So: ὀρθῶς γὰρ λέγεις.
Streps: ἐκεῖνο δ' ἦν ἂν "καρδόπη Κλεωνύμῃ".
So: ἔτι δέ γε περὶ τῶν ὀνομάτων μαθεῖν σε δεῖ,
ἅττ' ἄρρεν' ἐστίν, ἅττα δ' αὐτῶν θήλεα.
Streps: ἀλλ' οἶδ' ἔγωγ' ἃ θήλε' ἐστίν.
So: εἰπὲ δή.
Streps: Λύσιλλα, Φίλιννα, Κλειταγόρα, Δημητρία.
So: ἄρρενα δὲ ποῖα τῶν ὀνομάτων;
Streps: μυρία.
Φιλόξενος, Μελησίας, Ἀμυνίας.
So: ἀλλ' ὦ πόνηρε, ταῦτά γ' ἔστ' οὐκ ἄρρενα.
Streps: οὐκ ἄρρεν' ὑμῖν ἐστιν;
So: οὐδαμῶς γ', ἐπεὶ
πῶς ἂν καλέσειας ἐντυχὼν Ἀμυνίᾳ;
Streps: ὅπως ἄν; ὡδί·"δεῦρο δεῦρ', Ἀμυνία".
So: ὁρᾷς; γυναῖκα τὴν Ἀμυνίαν καλεῖς.
Streps: οὔκουν δικαίως, ἥτις οὐ στρατεύεται;
ἀτὰρ τί ταῦθ' ἃ πάντες ἴσμεν μανθάνω;
So: οὐδέν, μὰ Δί'.

So: But there are other things you must learn before that,
such as which of the quadrupeds are properly called masculine.
Streps: Well, I know which are masculine, if I've not lost my wits:
ram, he-goat, bull, dog, fowl.
⟨So:⟩ ⟨And feminine are –?⟩
⟨Streps:⟩ ⟨Ewe, she-goat, cow, bitch, fowl.⟩
So: Do you see what you're doing? You call the female
'fowl', and the male as well you call the same thing.

Streps: How do you mean, please?
So: How do I mean? 'Fowl' and 'fowl'.
Streps: By Poseidon, so I do. What should I really call it?
So: 'Fowless', and the other one 'fowler'.
Streps: 'Fowless'? By Air, that's a good one.
In fact, for that piece of instruction alone
I'll fill the whole surface of your *cardopus* with groats.
So: There you go again; that's another one. You speak of
a *cardopus*, calling it masculine when it's feminine.
Streps: What do you mean?
I call a *cardopus* masculine?
So: You certainly do,
just like *Cleonymus.*
Streps: In what way? Tell me.
So: You give the same treatment to *cardopus* that is given to *Cleonymus.*
Streps: But my good man, Cleonymus never had a *cardopus* at all –
the kneading he did was done in a round mortar.
What should I call it in future, though?
So: What should you call it?
Cardopé, just as you say *Sostraté.*
Streps: *Cardopé*, feminine?
So: That's correct.
Streps: And what I said would then have been 'Cleonymé never had a *cardopé*'?
So: But you still have to learn about names,
which of them are masculine and which feminine.
Streps: Well, I know which are feminine.
So: Tell me then.
Streps: Lysilla, Philinna, Cleitagora, Demetria.
So: And which names are masculine?
Streps: Loads of them –
Philoxenus, Melesias, Amynias.
So: But those aren't masculine, you fool.
Streps: You don't regard them as masculine?
So: Certainly not; consider,
if you happened to see Amynias, how would you call him?
Streps: How? Like this: 'Come here, come here, Amynia'.
So: Do you see? You're calling *her* a woman, 'Amynia'.
Streps: And doesn't she deserve it, for not doing her military service?
But why should I be learning these things, which we all know?
So: Never mind that now. (Aristophanes, *Clouds* 658–694)[2]

Here the expertise is based on observation of regularities in language and the construction of rules based on those regularities; those rules are in turn used to produce arguments that what ordinary people do is wrong, arguments that win praise and admiration. There must be some resemblance between this passage and debates actually occurring in fifth-century Athens, as otherwise the parody would not have been funny. At a minimum this resemblance probably includes the expert's success in con-

2 Translation from Sommerstein (1991) 71–75.

vincing others of the validity and value of his linguistic knowledge (although part of the joke here may be on Strepsiades for being so easily impressed, the whole set of arguments would not have been worth parodying if they had not been gaining any traction at all in real life) as well as the basis of the expertise itself: the construction of rules based on observed regularities in language.

At the same time the parody is unlikely to be accurate in every detail, and indeed one major inaccuracy is apparent: nothing else we know about Socrates suggests that he was a linguistic expert. Indeed, a claim of expert status in any area would be a major contradiction to Plato's portrayal of Socrates as someone who presented himself as knowing nothing. While allowing that Plato's portrayal is also unlikely to be completely accurate, most scholars think that the subject of this parody is probably not Socrates at all. The *Clouds* mocks sophists in general and uses the name 'Socrates' for a composite character made up of features of different individuals, and here the sophists being parodied are likely to be Prodicus and/or Protagoras.[3]

The basis of expertise portrayed in this passage is part of a larger pattern of how early Greek philosophy worked: the philosophers discovered previously unnoticed regular patterns in mathematics, astronomy, and other aspects of the natural world, and in many cases they were able to use those patterns to gain real information that non-experts did not have. Given the largely regular nature of language, it would have been surprising if early Greek philosophers had not noticed the regular patterns of Greek and played around with the irregularities.

Another, much later, parody suggests the same basis for linguistic expertise. This comes from the Roman-period Sceptic philosopher Sextus Empiricus, who was opposed on principle to experts of any sort: Sextus systematically attacked people who claimed to be experts, on the grounds that they were fakes. Significantly, linguistic expertise gets pride of place in his attacks, with 'Against the Grammarians' being the first book of a systematic demolition of experts (other targets are specialists in rhetoric, geometry, arithmetic, astrology, music, logic, physics, and ethics). This prominent position shows that in Sextus' day there must have been a body of linguistic experts large enough, and influential enough, to make good targets – and they must have used arguments broadly similar to the ones he tries to refute. The crucial elements of his argument can be seen in passage 2.

3 See e.g., Sommerstein (1991) 196 and Dover (1968) 182. This passage is included in collections of imitations of Protagoras by Diels/Kranz (1952) §80 C 3 and Capizzi (1955) 214–215.

2) Εἰ ἔστι τις τέχνη περὶ ἑλληνισμοῦ

Ὅτι μὲν δεῖ τινα φειδὼ ποιεῖσθαι τῆς περὶ τὰς διαλέκτους καθαριότητος, αὐτόθεν συμφανές· ὅ τε γὰρ ἑκάστοτε βαρβαρίζων καὶ σολοικίζων ὡς ἀπαίδευτος χλευάζεται, ὅ τε ἑλληνίζων ἱκανός ἐστι πρὸς τὸ σαφῶς ἅμα καὶ ἀκριβῶς παραστῆσαι τὰ νοηθέντα τῶν πραγμάτων. ἤδη δὲ τοῦ ἑλληνισμοῦ δύο εἰσὶ διαφοραί· ὃς μὲν γάρ ἐστι κεχωρισμένος τῆς κοινῆς ἡμῶν συνηθείας καὶ κατὰ γραμματικὴν ἀναλογίαν δοκεῖ προκόπτειν, ὃς δὲ κατὰ τὴν ἑκάστου τῶν Ἑλλήνων συνήθειαν ἐκ παραπλασμοῦ καὶ τῆς ἐν ταῖς ὁμιλίαις παρατηρήσεως ἀναγόμενος. οἷον ὁ μὲν τῆς Ζεύς ὀρθῆς πτώσεως τὰς πλαγίους σχηματίζων τοῦ Ζεός τῷ Ζεΐ τὸν Ζέα κατὰ τὸν πρότερον τοῦ ἑλληνισμοῦ χαρακτῆρα διαλέλεκται, ὁ δὲ ἀφελῶς τοῦ Ζηνός λέγων καὶ τῷ Ζηνὶ καὶ ⟨τὸν⟩ Ζῆνα κατὰ τὸν δεύτερον καὶ συνηθέστερον ἡμῖν. ... Ἄρτι μὲν οὖν ἐκ τῆς πρὸς τοὺς γραμματικοὺς ἀκολουθίας συνῆκται τὸ παρέλκειν μὲν τὴν ἀναλογίαν πρὸς ἑλληνισμόν, εὐχρηστεῖν δὲ τὴν τῆς συνηθείας παρατήρησιν.

Whether there is an expertise about Hellenism

That one must take some care over purity of one's language is immediately obvious, since one who constantly commits barbarisms and solecisms is ridiculed as ignorant, while one who speaks good Greek is able to express clearly and accurately what he has in mind. Now there are two kinds of Hellenism, for one form is divorced from our common usage and seems to proceed according to grammatical analogy, while the other form accords with the usage of each of the Greeks and advances by assimilation and observation in conversation. Thus one who forms the oblique cases of the nominative Ζεύς as Ζεός, Ζεΐ, Ζέα has spoken according to the first kind of Hellenism, while one who simply says Ζηνός, Ζηνί, and Ζῆνα has spoken according to the second kind of Hellenism, to which we are more used. ... So it has just been deduced from the consequences of the grammarians' own argument that analogy is superfluous for Hellenism, while the observation of common usage is most useful. (Sextus Empiricus, *Adversus mathematicos* 1.176–77, 209)[4]

The basis for claims of linguistic expertise here seems to be fundamentally the same as it was in Aristophanes' day: knowledge of rules based on analogy allows people to claim expert status. Those rules have been refined in the intervening centuries, and terminology has been invented to frame them (most notably ἀναλογία itself), but the basic principle remains the same: linguistic experts are parodied as rejecting common usage in favour of weird creations that no-one actually uses but that ought to exist if the Greek language were more regular than it is. Again, the parody relies on the fact that the linguists' analogical rules produce results that seem preposterous to ordinary native speakers, but this time there is an additional complication.

The example Sextus chooses, the paradigm of Ζεύς, is a complex one. A modern Hellenist might not see his 'regular' declension Ζεός, Ζεΐ, Ζέα as the expected one here; modern grammars usually give the paradigm for nouns in -ευς as βασιλεύς, βασιλέως, βασιλεῖ, βασιλέα,[5] so we might expect a 'regularised' Ζεύς to decline Ζέως, Ζεῖ, Ζέα. But these forms are the Attic ones with quantitative metathesis in the genitive singular; most other dialects had other forms, and when ancient gram-

4 Translation from Blank (1998) 37–38, 42.

5 See Smyth (1956) §275; Goodwin (1894) §263. For ease of comparison with Sextus' argument, I give all paradigms in the ancient order: nominative, genitive, dative, accusative (followed by vocative if included).

marians discussed the declension of Ζεύς they largely agreed with Sextus about the way it would have declined if it had been regular. Passage 3, from the Byzantine writer Choeroboscus, is thought to come ultimately from Herodian, who lived in the second century CE.[6]

3) Εἰ ἄρα οὖν τὸ Ζεύς εἰς -ς λήγει καὶ μονοσύλλαβόν ἐστι καὶ ὀξύτονον καὶ διὰ καθαροῦ τοῦ -ος κλίνεται, ὀφείλει Ζευός εἶναι κατὰ τὴν γενικὴν διὰ τῆς ευ διφθόγγου, ἵνα φυλάξῃ τὸν χρόνον τῆς εὐθείας ἐν τῇ γενικῇ· ἀλλ' ἐπειδὴ τὰ εἰς -ευς ἀποβάλλουσι τὸ υ κατὰ τὴν γενικήν, οἷον Πηλεύς Πηλέος, Ἀχιλλεύς Ἀχιλλέος, βασιλεύς βασιλέος, Θησεύς Θησέος, ... τούτου χάριν ἠναγκάζετο Ζεός εἶναι κατὰ τὴν γενικὴν κατὰ ἀποβολὴν τοῦ υ· τούτων οὖν τῶν δύο κανόνων μαχομένων, καὶ τοῦ μὲν ἑνὸς ἀπαιτοῦντος φυλάττεσθαι τὸ υ κατὰ τὴν γενικὴν διὰ τὸ φυλαχθῆναι τὸν χρόνον τῆς εὐθείας, τοῦ δὲ ἑτέρου ἀπαιτοῦντος ἀποβάλλεσθαι τὸ υ κατὰ τὴν γενικὴν τῷ λόγῳ τῶν εἰς -ευς, ἀπέλειψεν ἡ κοινὴ κλίσις καὶ παρεισῆλθεν ἡ τῶν Βοιωτῶν κλίσις, οἷον Διός Διί Δία, ὥσπερ Ἀχιλλίος Ἀχιλλΐϊ Ἀχιλλία.

So therefore if Ζεύς ends in -ς and is a monosyllable and oxytone and is declined with a pure [i.e., *preceded by a vowel*] -ος ending, it should be Ζευός in the genitive with the diphthong ευ, so that it should keep the [*vowel*] length of the nominative in the genitive; but since nouns in -ευς lose the υ in the genitive, like Πηλεύς gen. Πηλέος, Ἀχιλλεύς gen. Ἀχιλλέος, βασιλεύς gen. βασιλέος, Θησεύς gen. Θησέος ... for this reason it should have been Ζεός in the genitive with loss of the υ. So since these two rules conflict, and one demands that it keep the υ in the genitive in order to keep the [*vowel*] length of the nominative, and the other demands that it lose the υ in the genitive on the principle of nouns in -ευς, the common declension has failed and the Boeotian declension has been introduced, thus Διός, Διί, Δία, like [*Boeotian*] Ἀχιλλίος, Ἀχιλλΐϊ, Ἀχιλλία. (Choeroboscus' commentary on Theodosius' *Canons*, *Grammatici Graeci* IV.I 214.15–20, 23–29)[7]

Here Herodian is not suggesting that Ζεύς should be declined like βασιλεύς – he is perfectly clear that its actual oblique forms are Διός, Διί, and Δία –, but he does maintain that if Ζεύς were regular, it would have to decline either like βασιλεύς or according to another rule that it also does not follow. Theodosius (fourth/fifth century CE) has a similar discussion in the *Canons*, a set of prescriptive rules for the inflection of Greek words; he treats nouns in -ευς in canon 10.

4) Κανών ι΄
Ἑνικά. Ὁ Πηλεύς τοῦ Πηλέος, Πηλέως δὲ Ἀττικῶς· τὰ εἰς -ευς λήγοντα διὰ τοῦ -εος κλίνεται, Οἰνέος Τυδέος Πηλέος· Ἀττικοὶ δὲ τῆς γενικῆς ταύτης τρέπουσιν τὸ ο εἰς ω, Πηλέως Οἰνέως Τυδέως. τῷ Πηλέϊ καὶ Πηλεῖ κατὰ συναίρεσιν, τὸν Πηλέα, ὦ Πηλεῦ· εἴπομεν ὡς τὰ εἰς -ς λήγοντα μετὰ διφθόγγου ἀποβολῇ τοῦ -ς ποιεῖ τὴν κλητικήν, ὁ Ζεύς ὦ Ζεῦ.

6 See Lentz's edition of Herodian (*Grammatici Graeci* III.II 674.25–40), where this passage is given as part of Herodian's Περὶ κλίσεως ὀνομάτων. The attribution is uncertain, as with most material attributed to Herodian; see Dyck (1993) and Dickey (2014). For additional discussions of Herodian's text and issues of scholarship, see the contribution by Stephanie Roussou in this volume.
7 Translations are my own unless otherwise noted.

Rule 10
Singular. Nominative Πηλεύς, genitive Πηλέος, but Πηλέως in Attic. Nouns ending in -ευς are declined in -εος, as Οἰνέος [*from* Οἰνεύς], Τυδέος [*from* Τυδεύς], and Πηλέος, but Attic speakers turn the ο of this genitive to ω, [*producing the genitives*] Πηλέως, Οἰνέως, and Τυδέως. Datives Πηλέϊ and Πηλεῖ with contraction; accusative Πηλέα, vocative Πηλεῦ. We said that nouns ending in -ς preceded by a diphthong make the vocative by dropping the -ς, as nominative Ζεύς, vocative Ζεῦ. (*Grammatici Graeci* IV.I 12.5 – 11; cf. Blank [1998] 209)

Here the Atticising paradigm that we normally learn is given, but only as an alternative to the one with genitive in -έος. Theodosius' treatment of Ζεύς is, while not actually inaccurate, highly misleading for a student: by including Ζεύς in this discussion of regular nouns in -ευς, and failing to point out that it does not in fact follow the rules given except in the one case mentioned, Theodosius strongly implies that it follows the rules. No doubt Theodosius knew perfectly well that it does not, but nevertheless if presentations like this were to be found in the works of genuine linguistic experts, Sextus' complaint about the ridiculous forms their rules would produce is not entirely unjustified.

Of course, neither of these passages predates Sextus, nor indeed do any of the other surviving discussions of the declension of Ζεύς.[8] But given the poor survival rate of early grammatical literature we would not expect to have such discussions from before the second century CE even if they had been common; most likely the irregularity of Ζεύς had already been pointed out long before Sextus' day. Sextus made a point of reading works of the experts he attacked, using their techniques of argumentation, and referring to points they debated, in order to illustrate his proficiency in the subject and therefore his authority to point out its drawbacks.[9] It would therefore be entirely in keeping with his method to use an example from actual grammatical discussions here.

Therefore Sextus' first paradigm of Ζεύς actually makes perfect sense: it is what grammarians said that paradigm would have looked like if it had been regular. More surprising is the declension Sextus prefers instead. Ζηνός, Ζηνί, and Ζῆνα, though well attested (e.g., in Homer), were strictly poetic, and the normal oblique forms of Ζεύς were unquestionably Διός, Διί, and Δία. Sextus certainly knew this, for elsewhere he consistently employs the normal forms:[10] why does he not give those forms here? Two answers are possible: the Ζηνός paradigm may have been preferred because it began with the same letter as the nominative Ζεύς and therefore looked

8 These include Choeroboscus, *Epimerismi in Psalmos* p. 72.17 Gaisford; Sophronius, *Excerpta ex Joannis Characis commentariis in Theodosii Alexandrini canones* (*Grammatic Graeci* IV.II 390.30); *Epimerismi Homerici* ζ 3.6 Dyck; *Etymologicum Gudianum* ζ p. 230.4 Sturz; etc. They appear all to be related to one another and to Herodian's and Theodosius' discussions.

9 Sluiter (2000a).

10 Διός *Adversus mathematicos* 2.105, 5.29, 5.34, 5.36, 7.265, 9.36, 9.67, *Pyrrhoniae hypotyposes* 1.189; Διί *Adversus mathematicos* 1.290, *Pyrrhoniae hypotyposes* 3.224; Δία *Adversus mathematicos* 3.6, 3.104, 5.32, 6.3, 7.240, 7.445, 8.308, 8.375, 8.479, 9.115, *Pyrrhoniae hypotyposes* 1.50, 1.161, 2.141.

more regular (in which case Sextus would have been partially conceding the point that regularity makes correctness), or it may have been preferred precisely because it was poetic. We shall return to this point later.

Both Aristophanes and Sextus make it very clear that they, as non-experts, saw linguistic experts as characterised by knowledge of rules based on regularities, i.e. analogy. The writings of actual linguistic experts tend to confirm this picture. Herodian, for example, in his Περὶ μονήρους λέξεως 'On lexical singularity' (the only work of Herodian to survive intact and therefore the only one to preserve discussions of this type) presents analogy as the key to understanding Greek and portrays Herodian himself as the champion of analogy.[11] In his other, fragmentary works Herodian implicitly makes the same point by constantly using rules based on analogy to argue for the correctness or incorrectness of particular variants. Passage 5 is typical:

5) ἐπήτριμοι: ὁ Ἀσκαλωνίτης ὀξύνει ὡς πολεμικοί, οὐκ εὖ· τὰ γὰρ διὰ τοῦ -ιμος ὑπὲρ δύο συλλαβὰς τρίτην ἀπὸ τέλους ἔχει τὴν ὀξεῖαν, νόμιμος σπόριμος. οὕτως καὶ ἐπήτριμος.
Ἐπήτριμοι: Ptolemy of Ascalon gives this an oxytone accent, like πολεμικοί, but this isn't right; for words in -ιμος more than two syllables long have an acute on the third syllable from the end, like νόμιμος and σπόριμος. Thus also ἐπήτριμος. (A-scholion to *Iliad* 18.211b)[12]

Herodian's father Apollonius Dyscolus, who also wrote in the second century CE, states that his goal in formulating the rules of syntax is to show the extent of regularity in Greek, as passage 6 illustrates.

6) εἰς γὰρ τὸ τοιοῦτο τὰ τῆς συντάξεως ἀπεδείχθη, ἵνα καὶ τὰ λεληθότα τῶν ὑπερβατῶν, ὄντα καὶ κατὰ τὸ σύνηθες, ἔχηται τοῦ ἀκολούθου λόγου, καὶ μὴ μόνον τὰ παρὰ ποιηταῖς ὡς ἐξαίρετά τις ὑπολαμβάνοι.
For the rules of syntax have been demonstrated for just this purpose, so that even the unnoticed disturbances, which exist even in normal usage, may participate in the rule of regularity, and lest someone assume that only those disturbances used by the poets are exceptional. (Apollonius Dyscolus, *Syntax* 183.14–16 = II.77)[13]

Most of the time, the correctness the grammarians cared about was a correctness of authenticity in the language of Homer or another early author; for example, Herodian's point in passage 5 is that in reciting the *Iliad* one should use a particular accentuation for ἐπήτριμοι, an obscure word that was no longer part of the spoken language. Had ἐπήτριμοι still been in current usage, there would probably not have been a question about how to accent it, since native speakers would simply have known which syllable bore the accent.[14] This correctness is different from the kind

11 Suiter (2011).
12 Text follows Hartmut Erbse (1969–1988). This is given as a fragment of Herodian's Περὶ Ἰλιακῆς προσῳδίας by Lentz in *Grammatici Graeci* III.II 107.13–15.
13 Translation from Blank (1994) 158–159.
14 Although there were some changes in the nature of the Greek accent after the Classical period – most significantly, the distinction between acute and circumflex ceased to be audible – the position

of correctness envisioned by Sextus or Aristophanes in their parodies of linguistic experts: the parodies depict experts making prescriptive judgements about contemporary usage. In general, the surviving fragments of the Alexandrian grammarians and other early linguistic experts do not show much evidence of interest in prescribing contemporary usage, but there are a few passages in which ancient grammarians do look as though they are being prescriptive – and using analogy to support their judgements. In passage 7, for example, Apollonius seems to use analogy to argue that some Greek speakers are wrong in their pronunciation of ἀνεκάς.

7) Τὰ εἰς -ας λήγοντα ὀξύνονται ... οὕτως ἔχει καὶ τὸ ἐντυπάς, ἑκάς, ἀνεκάς (ὅπερ Ἀττικοὶ οὐ δεόντως ἀναβιβάζουσιν, ὡς καὶ ἐν ἑτέροις ἐπιρρήμασι, χάριέν φασι καὶ ἄληθες, καθὼς δείκνυμεν καὶ ἐπὶ τοῦ σφόδρα).—σημειωτέον οὖν τὸ πέλας.
Words ending in -ας are oxytone ... thus also ἐντυπάς, ἑκάς, and ἀνεκάς (on which the Attic speakers incorrectly retract the accent, as they also do in other adverbs, for they say χάριεν and ἄληθες, as I show also in my discussion of σφόδρα); but πέλας has to be taken as an exception. (Apollonius, Dyscolus, *De adverbiis* 160.19 – 23)

In this passage Apollonius is not in fact doing what the parodies say that linguistic experts do: he is not using a rule derived from analogy to argue that what everyone does is wrong. The accent of πέλας does not fit the analogical rule, and Apollonius simply labels it as an exception rather than prescribing the analogically more regular accentuation πελάς. But what is going on with ἀνεκάς? The obvious interpretation of Apollonius' words here is that when one group of Greek speakers has one accentuation and another group has another accentuation, only one of them can be right, and analogy will show which one that is. This may not be exactly what Apollonius really meant, for like all Greek grammarians he was well aware of the differences between Greek dialects. An Attic, Doric, or Aeolic feature not shared by the other dialects was not normally considered 'right' or 'wrong' in an abstract sense, but simply seen as a dialect feature. It would be appropriate in text using that dialect, and inappropriate in text using a different dialect. So perhaps Apollonius' οὐ δεόντως did not really mean 'incorrectly'; perhaps what he really meant was something closer to 'exceptionally'.[15] Nevertheless, a reader could not be blamed for interpreting this passage as saying that the analogical rule shows that the Attic accentuation is wrong; statements like this probably helped create the image of a linguistic expert parodied by Sextus.

Passage 8 is very similar: it appears to use analogical rules to decide which of two groups of Greek speakers is right, though again this may not be quite what the writer intended to convey. This passage, however, is considerably earlier, coming

of the accent was not normally affected. Most ancient words that survive into modern Greek still have the accent on the same syllable as in the fifth century BCE. Native Greek speakers of all periods, therefore, rarely had difficulties knowing which syllable to accent when using ancient words that remained part of the living language.

15 I owe this idea to Philomen Probert (personal communication).

from the first century BCE; as such it demonstrates that this type of statement was not new in the second century CE, but based on a long tradition.[16]

8) ἔδει δὲ καὶ παρὰ τὸ κλέπτης καὶ ψευδὴς μηδ' ὅλως σχηματίζεσθαι συγκριτικά. οἱ δ' Ἀττικοὶ κλεπτίστατον καὶ ψευδίστατον εἰπόντες τὰ συγκριτικὰ ἥμαρτον· ἀντὶ γὰρ τοῦ κλεπτέστατον καὶ ψευδέστατον.
And one shouldn't have formed comparatives at all to κλέπτης and ψευδής. But the Attic speakers, who said κλεπτίστατον and ψευδίστατον, were wrong about the comparatives. For they should have been κλεπτέστατον and ψευδέστατον. (Philoxenus fr. 339 Theodoridis, quoted in *Etymologicum Genuinum* s.v. ἀνιηρέστατον)

It is notable that in both passage 7 and passage 8 the usage apparently characterised as not being right is that of Attic speakers. In passage 9 we get a different kind of analogical explanation of an Attic variant: here Attic is clearly stated to have its own, separate regularities, making it different from but not necessarily better or worse than the non-Attic forms. Unfortunately, this passage is hard to date; at least some of the basic ideas evidently come from the first-century BCE scholar Trypho, but they were reformulated in the early second century CE and then epitomised under the name 'Ammonius'.

9) εἰ δὴ τὰ τοῦ τύπου κεκοινώνηκεν, ἄμφω ὀφείλει καὶ τῶν αὐτῶν τόνων μετέχειν, ὅτι ὀξυτονεῖν ἀπαιτεῖ ὁ λόγος βίᾳ ἀνάγκης. πᾶν γὰρ παρώνυμον εἰς -ρος λῆγον παρασχηματιζόμενον τοῖς γένεσιν ὀξυτονεῖται· οἷον κάματος καματηρός, ὄλισθος ὀλισθηρός, μέλι μελιτηρός, τόλμα τολμηρός, νόσος νοσηρός· εἰ δὴ πόνος καὶ μόχθος τὰ πρωτότυπα, πονηρὸς καὶ μοχθηρὸς ῥητέον ὀξυτόνως. εἰ δ' οἱ Ἀττικοὶ βαρυτονοῦσιν, οὐ θαυμαστόν ἐστι· χαίρουσι γὰρ τῇ βαρύτητι. ἄδελφε γοῦν λέγουσι τὴν πρώτην ὀξυτονοῦντες, ὡς ἄπελθε, φησὶν ὁ Τρύφων παρατιθέμενος Φιλήμονα τὸν ἀγξωνέα, ὡς Θεττάλην ὡς Μυρτάλην· καὶ χάριεν, τὴν πρώτην συλλαβὴν ὀξυτονοῦντες.
So if they share the same original form, they should also both share the same accents; the argument demands by necessity that they be oxytone. For every derivative ending in -ρος formed similarly to its relatives is oxytone: thus καματηρός from κάματος, ὀλισθηρός from ὄλισθος, μελιτηρός from μέλι, τολμηρός from τόλμα, and νοσηρός from νόσος. Indeed if the base forms are πόνος and μόχθος, one has to say πονηρός and μοχθηρός with oxytone accent. And if the Attic speakers give them a recessive accent, it's not surprising, for they often give things recessive accents. For example they say ἄδελφε, putting an acute accent on the first syllable, like ἄπελθε (Trypho says, citing Philemon of Aixone), like Θεττάλην, like Μυρτάλην – and χάριεν, to which they give an acute on the first syllable. (Ammonius, ed. Nickau (1966), entry 405.8–19 = p. 116; cf. Trypho frag. 2.9)

The type of comment seen in examples 7–9, in which Attic has a status no higher than other dialects, reflects a fundamentally pre-Atticist viewpoint. Once Classical Attic was prescribed as the standard for judging contemporary written Greek,

16 Although generally speaking direct evidence for the reasoning of scholars earlier than the first century BCE is non-existent, note Ax's discussion of the possibility that Aristophanes of Byzantium was an 'Analogist' (Ax 1990).

'Attic' became virtually synonymous with 'correct', making statements such as the ones just quoted difficult for later readers to understand.[17]

Linguistic expertise based on knowledge of the Attic literary canon

Analogy was not the only criterion on which linguistic expertise could be based. Although the famous feud between Analogy and Anomaly is now thought never to have existed,[18] and anomaly would in any case have been very difficult to use as the basis for a claim of either correctness or expertise, a number of genuine criteria have been identified.[19] Only one of these, however, comes close to rivalling analogy's status as the basis of linguistic expertise, and that is attestation in the Attic literary canon. Experts relying on this criterion are parodied even more often than ones using analogy, and a number of these parodies can be found in the second-century CE writer Lucian. Lucian was of course an accomplished Atticist himself, but he vigorously attacked rival speakers both for excessive archaism and for making mistakes. When in the course of such attacks Lucian explains why something should be considered right or wrong, he never uses the type of analogical arguments we have so far seen; instead he bases claims to correctness on attestation in classical authors. This is what linguistic Atticism was fundamentally about: if a word or usage could be found in a good Classical or Archaic author, it was right, and if not, it was wrong. Now of course not all good Classical authors wrote in what we would call the Attic dialect, but during the Roman period the term 'Attic' came to be extended to include all Classical and Archaic authors whose style one might want to imitate, regardless of their actual dialect. This is how Homer came to be considered an Attic author.[20]

In passage 10, from the *Lexiphanes*, Lucian makes the point that a rival 'speaks like a foreigner': this man's excessive use of Attic archaisms makes his speech appear not Attic at all. Later in the same work Lucian shows, point by point, that his opponent misuses the obscure words of which he is so proud: the words themselves may be Classical Attic, but the meanings in which they are used are unattested in the Attic canon, and therefore wrong (passage 11).

17 Probert (2011).
18 For Varro's misrepresentation of grammatical debate and the non-existence of a school of Anomaly, see Blank (1994) 152–58. There are no appeals to Anomaly as a criterion of correctness in ancient linguistic discussions. See Siebenborn (1976).
19 Siebenborn (1976) lists the following in addition to analogy: literary tradition, usage, etymology, and dialect (that is, the comparison of forms in different dialects to shed light on underlying patterns), plus a few Latin criteria that are not attested in Greek.
20 Cf. Dickey (2007) 98.

10) ἡμᾶς τοὺς νῦν προσομιλοῦντας καταλιπὼν πρὸ χιλίων ἐτῶν ἡμῖν διαλέγεται διαστρέφων τὴν γλῶτταν καὶ ταυτὶ τὰ ἀλλόκοτα συντιθεὶς καὶ σπουδὴν ποιούμενος ἐπ' αὐτοῖς, ὡς δή τι μέγα ὄν, εἴ τι ξενίζοι καὶ τὸ καθεστηκὸς νόμισμα τῆς φωνῆς παρακόπτοι.
Leaving behind us who are talking to him now, he talks to us like a thousand years ago, twisting up his tongue and putting together these monstrous things and being very serious about them, as if it were something great if he should speak like a foreigner and debase the established currency of our language. (Lucian, *Lexiphanes* 20)

11) οἷον πρῴην τὸν θυμάλωπα οὐδὲ εἰδὼς ὅ τι σημαίνει, ἀπέρριψας οὐδὲν ἐοικότα τῷ ὑποκειμένῳ. καὶ οἱ μὲν ἰδιῶται πάντες ἐτεθήπεσαν ὑπὸ τοῦ ξένου πληγέντες τὰ ὦτα, οἱ πεπαιδευμένοι δὲ ἐπ' ἀμφοτέροις, καὶ σοὶ καὶ τοῖς ἐπαινοῦσιν, ἐγέλων. τὸ δὲ πάντων καταγελαστότατον ἐκεῖνό ἐστιν, ὅτι ὑπεράττικος εἶναι ἀξιῶν καὶ τὴν φωνὴν εἰς τὸ ἀρχαιότατον ἀπηκριβωμένος τοιαῦτα ἔνια, μᾶλλον δὲ τὰ πλεῖστα, ἐγκαταμιγνύεις τοῖς λόγοις ἃ μηδὲ παῖς ἄρτι μανθάνων ἀγνοήσειεν ἄν· οἷον ἐκεῖνα πῶς οἴει κατὰ γῆς δῦναι ηὐχόμην ἀκούων σου ἐπιδεικνυμένου, ὅτε χιτώνιον μὲν καὶ τὸ ἀνδρεῖον ᾤου λέγεσθαι, δουλάρια δὲ καὶ τοὺς ἄρρενας τῶν ἀκολούθων ἀπεκάλεις, ἃ τίς οὐκ οἶδεν ὅτι χιτώνιον μὲν γυναικὸς ἐσθής, δουλάρια δὲ τὰ θήλεα καλοῦσιν;
Just as recently you threw in the word θυμάλωπα, having no idea what it means, when it had no connection to what you were trying to say. And the laymen were all astonished when their ears were struck by this foreign term, but the educated men were laughing both at you and at those who praised you. And the most ridiculous thing of all is that when you want to be hyper-Attic, and have carefully modelled your speech into the most archaic form possible, you mix in with your words some, or rather many, mistakes that not even a child just beginning to study would make. For example, how I prayed to sink beneath the earth when I heard you revealing that you thought χιτώνιον also meant a male garment, and when you used the term δουλάρια also of male retainers – who does not know that a χιτώνιον is a woman's garment, and δουλάρια refers to females? (Lucian, *Lexiphanes* 24–25)

Athenaeus, who wrote in the late second and early third centuries CE, also provides a sustained parody of this sort of appeal to attestation in the person of a character nicknamed Κειτούκειτος 'is it attested or not?'.

12) ... Οὐλπιανὸς ὁ Τύριος, ὃς διὰ τὰς συνεχεῖς ζητήσεις, ἃς ἀνὰ πᾶσαν ὥραν ποιεῖται ἐν ταῖς ἀγυιαῖς, περιπάτοις, βιβλιοπωλείοις, βαλανείοις, ἔσχεν ὄνομα τοῦ κυρίου διασημότερον Κειτούκειτος. οὗτος ὁ ἀνὴρ νόμον εἶχεν ⟨ἴδιον⟩ μηδενὸς ἀποτρώγειν πρὶν εἰπεῖν "κεῖται ἢ οὐ κεῖται;" οἷον εἰ κεῖται ὥρα ἐπὶ τοῦ τῆς ἡμέρας μορίου, ⟨εἰ ὁ μέθυσος ἐπὶ ἀνδρός,⟩ εἰ ἡ μήτρα κεῖται ἐπὶ τοῦ ἐδωδίμου βρώματος, εἰ σύαγρος κεῖται τὸ σύνθετον ἐπὶ τοῦ συός.
... Ulpian the Tyrian, who on account of the constant questions that he asks at every hour in the streets, walkways, bookshops, and baths has acquired the name Κειτούκειτος, which is better known than his real one. This man had the idiosyncratic custom of not eating anything before asking, 'Is it attested or is it not attested?' For example, asking whether 'hour' is attested for a part of the day, whether 'drunk' is attested applied to a man, whether 'paunch' is attested for the edible food, whether the compound word 'wild-boar' is attested for a pig. (Athenaeus, *Deipnosophistae* 1.2d-e)

Like Lucian, Κειτούκειτος focusses on the attestation of usage rather than simply the attestation of vocabulary. The latter was comparatively easy to ascertain in the second century, when a plethora of Atticistic lexica meant that one only had to look words up in a dictionary to find their ages, but the former required real knowledge of the Classical canon and was therefore the area on which disputes about expertise primarily focussed. The word κεῖται 'is attested (in a canonical text)' was evidently a key technical term in this context, embodying the concept of attestation in the Classical canon. It is widely used by writers on language in the same sense as by Athenaeus' Κειτούκειτος, as illustrated by passages 13–15, which range in date from the second to the fifth century CE.

13) Ἐργοδότης οὐ κεῖται, τὸ δὲ ἐργοδοτεῖν παρά τινι τῶν νεωτέρων κωμῳδῶν, οἷς καὶ αὐτοῖς οὐ πειστέον.
Ἐργοδότης is not attested, and ἐργοδοτεῖν is attested only in one of the more recent comic poets, whom (as a group) one should not trust. (Phrynichus, *Eclogae* 322)

14) τὰ μέντοι ἀπὸ τοῦ ρ ἀρχόμενα συνάρχεται τοῖς ἰδίοις παρατατικοῖς, ῥαπίζω ἐρράπιζον ἐρράπικα, ἔρραπτον ἔρραφα· τὸ γὰρ ῥερυπωμένα κεῖται ποιητικῶς.
But the perfects that begin with ρ- have the same beginning as their own imperfects, as [*present*] ῥαπίζω, [*imperfect*] ἐρράπιζον, [*perfect*] ἐρράπικα, [*or present* ῥάπτω, *imperfect*] ἔρραπτον, [*perfect*] ἔρραφα; for [*perfect*] ῥερυπωμένα is attested [*only*] in poetry. (Theodosius, *Grammatici Graeci* IV.I 48.8)

15) μαλάχη· οὐ μολόχη· παρ' οὐδενὶ γὰρ κεῖται.
One should say μαλάχη, not μολόχη, for that is not attested in any canonical author. (Orus, fragment 95 in Alpers 1981)

There was, however, a major problem with using κεῖται in this fashion: this sense of κεῖται is itself post-classical. Indeed, it would not have taken any inspection of the Classical canon to realise that, for the Classical canon itself must by definition be a post-Classical concept, and the concept of attestation in the Classical canon cannot predate the concept of the canon. Many educated Greek speakers were well aware of this fact and therefore avoided the term; Lucian, for example, never uses κεῖται to mean 'attested in the canon'. Many others, however, used the term as freely as Athenaeus' Κειτούκειτος – and therefore slipped up, according to their own criteria, every time they did so. Athenaeus was no doubt aware of this ironic situation and expected his more discerning readers to realise that poor Κειτούκειτος revealed his ignorance every time he asked a learned question.

The earliest examples of κεῖται in this sense probably come from the first century CE, if in fact it is to be found in passages 16 and 17. Passage 16 is a fragment without context; passage 17 is one of a number of similar examples from Erotian's lexicon of Hippocratic words.

16) ἢν εἴπῃς παρ' ὅτῳ (τὸ δεῖνα) κεῖται, δώσω σοι χρυσέους δέκα στατῆρας.
If you can tell me in what author that is attested, I'll give you ten gold staters. (Heraclides Ponticus Junior, fragment 1 in Heitsch (1964) 41)

17) οἰκείης· δούλης, οἱ δὲ ἰδίας. κεῖται ἐν τῷ Περὶ φύσεως παιδίου ἡ λέξις.
'Οἰκείης means 'of a female slave', but others say that it means 'personal [*gen. sing. f.*]'. The word is attested in the work entitled 'On the nature of the child'. (Erotian 101.4 in Nachmanson 1918)

The first century is therefore likely to be the period in which the concept represented by κεῖται, attestation in the Classical canon, first gained traction as a criterion of expertise. This concept cannot, however, have grown very influential by the time of Sextus Empiricus, since his attack on linguistic expertise focusses exclusively on analogy and does not attack the Atticists – who, as both Lucian and Athenaeus show, made excellent targets for ridicule once their movement was well established.

The unseen role of vernacular Greek

We see, therefore, two shifts in Greek thinking about linguistic expertise. The first occurred in the fifth century BCE, when the concept of the linguistic expert first arose: at that point claims to linguistic expertise were based on observation of regularities and extrapolation of rules from them. The second shift occurred in the second century CE, when the defining characteristic of the linguistic expert ceased to be knowledge of analogy and began to be knowledge of the Attic canon instead. The grammarians themselves did not stop using analogy, since nothing ever really disappeared from the Greek grammatical tradition. But they and their analogical rules ceased to be seen by others as the paradigm of linguistic expertise. After Sextus Empiricus, laymen thinking about linguistic experts had a stereotype of an expert in the Classical Attic literary canon, rather than a stereotype of an expert in analogical rules. How did this change come about?

The shift is unlikely to have come from the experts themselves, not only because they had a well-established vested interest in continuity but also because expert status is conferred primarily by outsiders. It is not an individual's possession of special knowledge, but the recognition by others that that individual's knowledge is special, that makes someone an expert.[21]

21 There is of course the complication that in some circumstances a society assigns expert status in a particular area to a clearly specified group, which is entrusted with the responsibility for deciding whom to admit to their group. This is the case with, for example, the licensing of doctors and electricians today: only doctors can decide who counts as a doctor, and only electricians can decide who counts as an electrician. But it is society as a whole, not the experts themselves, that gives validity to those licensing decisions: if laymen (such as those making the laws that forbid practising medicine without a licence) did not care whether or not someone was licensed to practice medicine, such licensing would have little practical impact. Even today, many skilled professions are not regulated in the same way: in England, for example, anyone can claim to be a plumber, and any non-plumber can judge whether that claim is accurate by hiring the alleged plumber to fix a plumbing problem. In antiquity most fields of expertise operated as plumbing does today, for no bodies of experts were organised to the extent of having meaningful licensing powers. It is clear that claims to linguistic expertise were disputed by other experts – that is what is going on in Lucian's *Lexiphanes* – but without an appointed authority to judge such disputes, they could have no clear process of resolution. Ultimately each person, no matter how much or how little he or she knew about the subject, would have had to make up his or her own mind about whom to trust as an expert.

During the period between the fifth century BCE and the second century CE ordinary Greeks, those who were not linguistic experts, experienced a significant change in their relationship to the Greek language. In the Classical period Athenians simply spoke and wrote their native language and did not worry about linguistic rules unless they happened to encounter someone like Protagoras – and even then they do not appear to have changed their linguistic practice to fit his rules. But by the start of the second century CE even Sextus Empiricus, who did not think there was any such thing as an expert, had to begin his attack on grammar with an acknowledgement that speakers needed to take care to speak good Greek; those who did not would be ridiculed for ignorance (passage 2). The everyday spoken language had changed gradually over time, as living languages always do, and after more than half a millennium the accumulated changes were significant; meanwhile the written standard had changed much less. In consequence, even before the rise of Atticism a major gap had opened up between the Greek that speakers were supposed to produce and the Greek that they found it easy to produce.

Although these changes in Greek came in all aspects of the language, from phonetics through morphology and the lexicon to syntax, the morphological changes were particularly striking because of their consistent direction: they tended towards regularisation. The complex irregularities of Classical Greek inflection underwent wholesale simplification and regularisation, as evidenced by the usage in documentary papyri. For example, the aorist of δίδωμι changed from ἔδωκα, ἔδωκας, ἔδωκε, ἔδομεν, ἔδοτε, ἔδοσαν to ἔδωκα, ἔδωκας, ἔδωκε, ἐδώκαμεν, ἐδώκατε, ἐδώκασαν. The aorists of λείπω and ἄγω changed from ἔλιπον and ἤγαγον to ἔλειψα and ἦξα. The declension of χάρις changed from χάρις, χάριτος, χάριτι, χάριν to χάρις, χάριτος, χάριτι, χάριτα.[22]

Such changes would naturally have led ordinary Greek speakers to have less and less respect for analogy. Linguistic communities tend to consider 'good' the speech of the people who have high status in the community, and 'bad' the speech of the people who have low status. In the Roman empire there was a considerable connection between status and education, and therefore it was more often the low-status people who used the newer, more regular forms. Now the grammarians themselves were skilled in the application of analogy and would not have used it to justify new forms used by low-status members of the community as opposed to the older, less regular forms used by the more prestigious speakers. But whatever the experts actually did, laymen could see for themselves that the principle of analogy was not one that worked for what they wanted.

What they wanted, what they valued, was expertise that helped them with the linguistic challenges they faced. The first of these was to read and understand the Classical and Archaic texts taught in schools: children learned to read on Homer during the Roman empire, just as they had done in Classical Athens, but Homer had be-

22 See Gignac (1981) *passim*.

come considerably more difficult during the intervening centuries. The second challenge was to produce correct Greek themselves, in writing and under some circumstances also in speech. A person who knew how to read, understand, and produce the kind of Greek that the non-experts aspired to master was the type of person they would acknowledge to be a linguistic expert.

Indeed, it is possible that this attitude on the part of non-experts was one of the catalysts for the rise of Atticism. Seeing a social and economic opportunity in the growing demand for linguistic expertise based on knowledge of the Classical canon, educated Greek speakers marketed themselves as experts of this type. Competition for this expert status (as evidenced by the rivalries between Lucian and other contemporary speakers) rapidly developed, leading to a sort of arms race in linguistic archaism. Then the production of aids such as the Atticising lexica raised the standards still further by making it easier for anyone to produce superficially educated-sounding Greek without actually mastering the entire canon. That in turn led the competition to focus on ever-more-difficult linguistic features, such as the attestation of particular usages rather than of words.

This situation may also explain Sextus' bizarre choice of the poetic Ζηνός paradigm rather than the usual Διός one in passage 2. As noted earlier, it is possible that Sextus preferred Ζηνός precisely because it was *not* what people normally used, but rather a Homeric paradigm. At the beginning of the second century CE Homeric forms had a growing cachet, and the ability to produce them was no doubt seen as a desirable form of expertise. Sextus' point that one does not need linguistic experts if one just pays attention to usage may have been greatly strengthened by the implicit demonstration that by paying attention to (Homeric) usage one can, without relying on experts, produce a fancy Homeric paradigm of Ζεύς. The Atticists were clearly not a significant enough group of experts for Sextus to consider them worth attacking, but that does not mean that he was completely unaware of their movement. Perhaps he found an implicit appeal to their ideas useful in his attempts to discredit the established linguists. If this is what was going on, we can see in passage 2 an early skirmish in the ultimate triumph of Atticism over analogy.

R.M.A. Marshall
Suetonius the Bibliographer

Suetonius is famed as the author of the *Vitae Caesarum*; he is less-well known for his extensive (but fragmentary) collection of literary biographies, the *De uiris illustribus*.[1] Suetonius' strengths and weaknesses as a biographer (particularly of Caesars) have been extensively studied and need no restatement here. The focus of this inquiry is instead upon a feature common to almost all Suetonian biographies: the inventory of the subject's written works. The provision of bibliography is so fundamental to the genre of literary biography that the enumeration of an author's works is generally taken for granted in modern studies, and tacitly assumed to follow the practice of Callimachus and other Hellenistic pioneers of the genre. At the turn of the twentieth century, Friedrich Leo drew attention to the peculiar provision of bibliography in Suetonius' 'political' lives, and, in arguing for the existence of 'Peripatetic' and 'Alexandrian' traditions of biography, pointed to similarities between Suetonius' literary and imperial bibliographies as proof that the 'wrong' biographical model had been adopted for the latter. According to his reconstruction, Suetonius had written the lives of emperors—men of action—as though they were literary figures in the 'Alexandrian' mode.[2]

Leo's rigid conception of genre (and speculative reconstructions of biographical traditions) has long been discredited. But it is perhaps time to re-examine Suetonius' use of bibliography as a literary device pregnant with hitherto unacknowledged rhetorical possibilities. An examination of Suetonian bibliographical practice demonstrates that in our author's hands, a wooden adjunct to scholarship has been turned into a flexible, even artistic tool. Rather than a mere appendix to biography, bibliography is used with such structural and thematic sophistication in Suetonius' portraits that it is more appropriate to talk of his 'biblio-biographies', rather than use a term associated with tedious lists and catalogues.[3]

The present study serves as an introduction to Suetonius' bibliographical practice, concentrating on structural features such as his avoidance of lists, ordering of books, and suppression of titles for local, rhetorical goals. Following an investigation

My thanks to Dr Sean A. Adams for organising the conference and to the participants, many of whom gave valuable advice on this paper.

1 Or perhaps *Catalogus uirorum illustrium:* Brugnoli (1968) 39–60.
2 Leo (1901).
3 'Biblio-biography' is here used for the thematic and structural blending of biography and bibliography. By extension, works such as Augustine's *Retractationes* or Galen's *On his Own Books* and *On the Order of his Own Books* might be termed 'biblio-autobiography'. Cf. Swain (2006) 396.

R.M.A. Marshall, University of Glasgow

https://doi.org/10.1515/9783110660982-009

of the models available to Suetonius, I will confine myself to a discussion of bibliography as deployed in the 'literary' lives, specifically the biographies of the grammarians from the *De grammaticis et rhetoribus*, and to the richest of the separately transmitted lives, namely those of Terence, Horace, and Vergil.[4] In my conclusion, I will also provide some tentative thoughts as to why Rome produced a Suetonius rather than a Callimachus. It is hoped that this preparatory study will form a basis for future discussion of Suetonius' bibliographical practice in the *Vitae Caesarum* and handling of titulature.

De uiris illustribus: Models

Ironically for a biographer so interested in books, the chronology of Suetonius' works relative to his rise through the ranks of the imperial bureaucracy cannot be reconstructed. At least part of the *Vitae Caesarum* was dedicated to the praetorian prefect Septicius Clarus *ca.* 119 – 122 CE.[5] No definite dates are associated with the *De uiris illustribus*, though potential allusions to the surviving life of Vergil in Pliny (*Epist.* 5.10) may provide a *terminus ante quem* of 105 – 106 CE for (parts of?) the collection.[6] From an inscription preserved at Hippo, we know that Suetonius served in several priesthoods and as a juror (under Trajan), and capped his career with a trio of high offices at the imperial court: *a studiis*, *a bibliothecis*, and *ab epistulis*.[7] He was dismissed from the latter post by Hadrian in 122 CE.[8] In the following discussion, I place no stress on the fact that Suetonius was once official librarian at Rome (*a bibliothecis*), firstly because it is not clear when precisely this post was held (if the *De uiris illustribus* did exist in some form by 110 CE, the work was partly responsible for—rather than a product of—Suetonius' career advancement),[9] and secondly, because it is not entirely clear what competences, duties, or privileges this post actually required, entailed, or granted.[10] The conclusions of this study, in fact, have negative implications regarding the scholarly potentialities of this position at Rome.

4 As the lives of the rhetors share the chief bibliographical features of the lives included in the present study, I have not thought it necessary to try the reader's patience with further examples of Suetonius' artful (lack of) method. The shorter 'Suetonian' lives of Tibullus, Lucan, Persius, Pliny the Elder, and Passienus Crispus likewise have little additional value as illustrations and have been subject to textual disruption, including in many cases severe abridgement: see below.

5 Lyd. *Mag.* 2.6 with Syme (1980) 68 – 69. Lindsay (1994) argues for Suetonius' dismissal in 128 CE, but see Wardle (2002).

6 Power (2010).

7 Marec/Pflaum (1952).

8 SHA *Hadr.* 11.3 (for the date, see n. 5).

9 Power (2010) 157 – 59. For a tentative chronology of Suetonius' career, see Townend (1961). *A studiis* 114/5 CE, *a bibliothecis* 116/7 CE, *ab epistulis* from 118 CE his until fall from favour.

10 The only true guide is provided by the office-holders themselves, an eclectic mixture of scholars and bureaucrats: Bowie (2013) Houston (2014) 232 – 44. The position plainly did *not* grant Suetonius

All that remains of Suetonius' collection of literary lives are a handful of individual biographies belonging to Terence, Vergil, Horace, Tibullus, Persius, Lucan, Pliny the Elder, and Passienus Crispus, which survive because they were used to fill commentaries or introduce editions of their respective authors in later antiquity. In addition, an extensive fragment known as the *De grammaticis et rhetoribus* preserves the biographies of all nineteen grammarians included in the collection, and five of the sixteen lives of rhetoricians. Saint Jerome explicitly relied upon Suetonius (*Chron.* pr. p. 6 Helm) for the dates of numerous secular authors incorporated into his fourth-century *Chronicon*, as a comparison of the *De grammaticis et rhetoribus* with the entries in the latter work also demonstrates. Thanks to Jerome, we can restore around fifty additional names to Suetonius' collection of literary lives.[11]

Jerome also allows us to glimpse the predecessors named in Suetonius' lost preface:[12]

> hortaris me, Dexter, ut Tranquillum sequens ecclesiasticos scriptores in ordinem digeram, et quod ille in enumerandis gentilium litterarum uiris fecit illustribus, ego in nostris faciam, id est ut a passione Christi usque ad decimum quartum Theodosii imperatoris annum omnes qui de scripturis Sanctis memoriae aliquid tradiderunt tibi breuiter exponam. fecerunt quidem hoc idem apud Graecos Hermippus Peripateticus, Antigonus Carystius, Satyrus doctus uir et longe omnium doctissimus Aristoxenus musicus, apud Latinos autem Varro, Santra, Nepos, Hyginus et ad cuius nos exemplum prouocas Tranquillus.
>
> You urge me, Dexter, to run through the ecclesiastical authors in order following [*Suetonius*] Tranquillus and to do for our authors what he has done in enumerating the famous men of pagan literature, that is to say, that I should set out briefly for you all those who have committed to posterity something about the Holy Scripture from the death of Christ to the fourteenth year of the Emperor Theodosius. Among the Greeks Hermippus the Peripatetic, Antigonus of Karystus, Satyrus, a learned man, and by far the most learned of all of them, Aristoxenus the musician, have done something similar, likewise among the Romans Varro, Santra, Nepos, Hyginus, and [*Suetonius*] Tranquillus, whose example you set before us. (Jerome, *De vir. ill.* pr.1)

This list is not complete: elsewhere, Jerome tells a correspondent that his *De uiris illustribus* followed the example of *Tranquillum Graecumque Apollonium* ('[*Suetonius*] Tranquillus and the Greek Apollonius': *Epist.* 47.3), which suggests of this passage that other authorities may also have been named by Suetonius that are now lost to us. The surviving fragments of many of these authors do not allow us to recon-

privileged access to the correspondence e. g., of Augustus, which is already quoted in the lives of Vergil and Horace (see below), and by authors before Suetonius. See Wallace-Hadrill (1983) 91–96.

11 See Roth (1858); Reifferscheid (1860); Helm (1929).

12 I thus support the traditional interpretation of this passage against Power (2014) 401–402, who objects on the grounds that Suetonius avoids 'self-conscious or rhetorical prefaces, or any comments on his literary purpose' in his 'extant beginnings'. As all the openings of Suetonius' works are lost (including that of the *Vitae Caesarum*), the value of this observation is doubtful. Jerome's knowledge of Greek literature also undermines confidence in his ability to reconstruct this canon from stray hints in Suetonius: Courcelle (1968) 58–127.

struct how they handled bibliography, if they did in fact do so.[13] Those who fall into this unfortunate category are Satyrus,[14] Aristoxenus,[15] Santra,[16] and Hyginus.[17]

The remains of Antigonus, Varro, and Apollonius provide room for speculation, but can support few firm conclusions. Antigonus was active in the third century BCE and wrote biographies of philosophers.[18] One fragment certainly implies that Antigonus prepared bibliographical lists:[19]

> ἦν δέ, φησὶν ὁ Ἀντίγονος, καὶ φιλοπότης καὶ ἀπὸ τῶν φιλοσόφων εἰ σχολάζοι ποιήματα συνέγραφε· καὶ γὰρ καὶ ἔπη καὶ τραγῳδίας καὶ σατύρους (καὶ δράματα κωμικὰ τριάκοντα, τὰ δὲ τραγικὰ ἑξήκοντα) σίλλους τε καὶ κιναίδους. φέρεται δ' αὐτοῦ καὶ καταλογάδην βιβλία εἰς ἐπῶν τείνοντα μυριάδας δύο, ὧν καὶ Ἀντίγονος ὁ Καρύστιος μέμνηται, ἀναγεγραφὼς αὐτοῦ καὶ αὐτὸς τὸν βίον.
>
> He (sc. Timon) was, says Antigonus, fond of drinking and when he had leisure from philosophy he wrote poetry: epic and tragedy and satire plays (and thirty comedies; there were sixty tragedies), besides *silli* (i.e. lampoons) and homoerotica. It is said there are prose books of his of 20,000 lines, as Antigonus of Carystus notes, who also wrote his biography. (Diogenes Laertius, 9.110 – 11 = Antigonus F5 Dorandi)

Antigonus also provided similar stichometric information (i.e. line counts) for the works of Crantor and discussed their authorship,[20] noted that Menedemus wrote nothing,[21] and was interested in the circulation of Plato's books.[22]

While much of Varro's varied biographical work was probably of limited relevance for Suetonius' bibliographical practice,[23] he did exercise a lasting impact on

---

13 Though some commentators insist, given the context, that all the authorities named must have been authors of 'literary' biographies, the surviving evidence for their biographical interests does not necessarily support this assumption. Suetonius' apparent interest in philosophical biography (as represented e.g., by Apollonius), however, can be explained by the inclusion of a section *De philosophis* in his own collection (F81 – 89 Reifferscheid). See Brugnoli (1968) 57 – 60; Schmidt (1991); Schmidt (2000) 42.

14 Schon (2004).

15 Kaiser (2010).

16 Funaioli (1907) 384 – 99.

17 Cornell *et al.* (2013) no. 63 F1 – 15.

18 See Dorandi (1999).

19 But does this passage 1) represent an abridgement of some more detailed list or discussion, 2) accurately report the extent of the information provided by Antigonus, or 3) supplement Antigonus with a list of the specific poetic genres embraced by Timon from some other source?

20 Dorandi (1999) F11 A, 19 A.

21 Dorandi (1999) F25.

22 Dorandi (1999) F20 A, 39.

23 The following works contained biographical material, but surviving fragments give no indication of how, or whether, bibliography also featured. *Imagines/ Hebdomades:* Salvadore (1999) F106 – 15. *De Pompeio:* Jerome, *Epist.* 33 (bibliographical content limited by subject: Pompey was no author; despite Cic. *Brut.* 239, his speeches do not seem to have circulated in antiquity). *De uita sua:* Cornell *et al.* (2013) no. 52 F4 – 6, political / military memoires, not biblio-autobiography: Tatum (2011)

the shape of the Plautine corpus.[24] Varro, in arguing for the authenticity of twenty-one titles, gave his name to a canon (Gellius, *NA* 3.3.3) that has survived to this day. Suetonius' life of Plautus is lost excepting a solitary testimonium,[25] yet given the influence wielded by Varro over Plautus' transmission and reception, Suetonius' missing biography likely relied (at whatever remove) on Varro's work. Plautus, however, seems to have been the exception where Varro was concerned: no other author is the named subject of one of his treatises.

If the identity of Apollonius was more firmly established, he might also have proved of interest. Suetonius can hardly have meant Apollonius of Tyana,[26] but a certain Apollonius of Tyre drew up a 'table of the philosophers who followed Zeno and of their books' (Strabo, *Geogr.* 16.2.10: πίναξ τῶν ἀπὸ Ζήνωνος φιλοσόφων καὶ τῶν βιβλίων).[27] Though conjectured to be an important source for Diogenes Laertius, no certain fragments of this work survive.[28] If David Hahm is correct in suggesting that Apollodorus brought his study down to the first century BCE,[29] this is all the more reason for Suetonius to have shown an interest in the collection: many of the latter's Roman subjects will have sat at the feet of Greeks written about by Apollonius.

If many of the authorities named above prove disappointing, two—Hermippus and Nepos—are vital for contextualising Suetonius' own bibliographical practice. Each represents one of the two chief forms taken by ancient bibliography, forms I will label the 'pinacographic' and 'biblio-biographical'. Hermippus also serves to bridge the gap between Suetonius and Callimachus, the 'father' of Greek bibliography: Hermippus was Callimachus' sometime assistant and built his own works on his master's legacy. Accordingly, it is to Callimachus we must first turn.

Callimachus, active in the third century BCE in the Alexandria of the Ptolemies, was deeply engaged in the new Hellenistic encyclopaedism,[30] and besides works

181–82. *De familiis Troianis:* Cornell *et al.* (2013) no. 52 F3. *De poetis:* Funaioli (1907) 209–12 (F55–62). The latter may have been a literary history, but could also have been structured like Suetonius' surviving *De grammaticis et rhetoribus*, with potted literary history giving way to individual biographies. For a conjectural reconstruction of the work, see Dahlmann (1963). Note that the *legationum libri* (Jerome, *Epist.* 33) is likely an administrative treatise, not autobiographical. Cf. Dahlmann (1973) 15–16.

24 Presumably via his *De comoediis Plautinarum* and/or *Quaestiones Plautinarum* V.

25 Jerome, *Chron.* a.A. 1817 (= 200 BCE).

26 Miller (1895). As an exact contemporary of Suetonius, he is unlikely to have been cited as a model. In any case, the *Life of Pythagoras* attributed to him in later antiquity may be apocryphal: Bowie (1978) 1672 n.77.

27 von Arnim (1895) 146. Identification summarily dismissed by Leo (1902) 17 n. 1.

28 Hahm (1992). The bibliographies of the Stoics may have formed an appendix to Apollonius' *Concerning Zeno*, which is frequently cited for anecdotes concerning that philosopher's life: Diogenes Laertius, 7.1, 2, 6, 24, 28.

29 Hahm (1992) 4161 n.195.

30 On Callimachus' encyclopaedism, see Hatzimichali (2013). See also the discussions by Catherine Heszer and Myrto Hatzimichali in this volume.

such as his *Aetia*, a poetic compendium of antiquarian lore, wrote a 120-volume catalogue of the works held by the famous Library, the Πίνακες τῶν ἐν πασῇ παιδείᾳ διαλαμψάντων καὶ ὧν συνέγραψαν ('Tables of all those preeminent in literature and their writings').[31] This was not a catalogue, in the modern sense, of the library's actual holdings. The surviving fragments give no suggestion that the books themselves were catalogued as individual items, thus no indication was given as to whether the library held one or several copies of the *Iliad*, for instance, nor what might be the age, provenance, condition, or quality of any particular copy held of any particular work.[32] What the *Pinakes* did provide, however, was a list of the works of a particular author, either of all those known to have been written by him or her (a bibliography in the true sense),[33] or of all those held by the Library (what Blum, in arguing for this interpretation, styles a 'catalogue of works').[34] Such a distinction, however, is largely immaterial, as the Alexandrian Library's holdings were very nearly exhaustive.[35] Each author was included by Callimachus under one of a number of generic classifications, so 'Orators' or 'Miscellaneous Literature'.[36] The authors were apparently arranged alphabetically by initial letter,[37] and each entry began with a brief summary of the author's epithets, place of birth, life, intellectual connections, and death.[38] The catalogue of the author's works followed this introductory material. Titles were distributed according to groups or genres,[39] but the logic of their arrangement within these groupings (if, in fact, there was any) is unknown. The titles were often accompanied not only by their first lines, or 'incipits',[40] but in an even greater show of precision, with stichometric information as well (that is, with an indication of the length of the work based on standardised line units).[41] Notes on the authenticity of particular works were also provided.[42] In effect, the *Pinakes* of Callimachus constituted the first attempt to write a national bibliography: it was presented as authoritative and aimed at systematic coverage.[43]

Callimachus' pinacographic work inspired imitators and continuators, including Hermippus the Peripatetic, named as a model in Jerome's (Suetonius') preface. Her-

---

31 *Suda* K 227. Fragments: Pfeiffer (1949) F429 – 453; *P.Oxy.* XXIII 2368.16 (published after Pfeiffer's edition). Major studies: Schmidt (1922); Regenbogen (1950) cols. 1418 – 23; Pfeiffer (1968) 126 – 30; Blum (1991).

32 Blum (1991) 159 – 60. For examples of ancient library catalogues, see Houston (2014) 39 – 86.

33 So Regenbogen (1950) col. 1420; Pfeiffer (1968) 128.

34 Blum (1991) 160, 226 – 39. Similar conclusion reached by Schmidt (1922) 99.

35 Blum (1991) 238 – 39.

36 Pfeiffer (1949) F430 – 32 (orators) 434 – 35 (misc.).

37 Pfeiffer (1949) F435.

38 Pfeiffer (1949) F429, 430, 438.

39 *P.Oxy.* XXIII 2368.16, cf. Pfeiffer (1949) F450.

40 Pfeiffer (1949) F433 – 34, 443 (cf. F436).

41 Pfeiffer (1949) F433 – 34.

42 Pfeiffer (1949) F442, 445 – 7, 449 (cf. F437, 444, 451).

43 Blum (1991) 238 – 39.

mippus served as Callimachus' research assistant,[44] and wrote, in effect, a popular supplement to the *Pinakes*, concentrating on certain individuals or groups (e.g., lawgivers, the philosophers Aristotle and Theophrastus, the rhetors Gorgias and Isocrates, and the latter's pupils) and greatly expanded the biographical component of their entries.[45] He retained (or adapted?) the dry bibliographical lists of his model, Callimachus,[46] and again concerned himself with questions of authenticity and ascription.[47] The result must have looked, at least in outline, a great deal like the lives of Diogenes Laertius or the *On Dinarchus* of Dionysius of Halicarnassus;[48] namely, a work arranged by author entries with lists of an author's titles appended to, or inserted within, their individual biographies. Erudition and entertainment, as in so much of the Alexandrian literary output, went hand in hand: Hermippus had a taste for the sensational and a particular fixation with remarkable deaths,[49] but he retained the 'pinacographic' book list as a vital component of his programme.

While pinacographical bibliography has received some attention from critics,[50] the alternative format adopted for bibliography in antiquity—the 'biblio-biographical'—seems not to have been described or studied on its own terms. It is instead treated as a secondary form, assumed to merely summarise pre-existing pinacographical bibliographies.[51] This may or may not hold true in the Greek world, but such assumptions are unwarranted at Rome, where many authors' written works must have been first enumerated in biblio-biographical format (of the individuals discussed below, Atticus and Suetonius' grammarians clearly fall into this category).

The biblio-biographical form is conveniently represented by Cornelius Nepos, the only author named by Jerome we have yet to consider. Rather than provide bare lists of titles, or a catalogue of entries following some standard format (e.g., title, first line, stichometric tally, note on authenticity if required),[52] 'biblio-biography' treats an author's corpus as a demonstration or expression of his character and/or significant life experiences, hanging stories and observations upon the individual titles.

This possibility ultimately stems from ancient biography's reliance on the fallacy that a subject's life can be reconstructed from his works.[53] For instance, Suetonius'

---

44 Bollansée (1999) 1–7.
45 Hermippus, *FGrH* 1026 F1–54.
46 Hermippus, *FGrH* 1026 T20, F9, 37, 38, 44a (F89?). See: Regenbogen (1950) cols. 1424–25; Wehrli (1974) 78–79; Blum (1991) 188–89; Bollansée (1999) 163–82.
47 Hermippus, *FGrH* 1026 F46b, 67.
48 See Bollansée (1999) 180. Dionysius complains that he was compelled to write because of the shortcomings of Callimachus' treatment: Dion. Hal. *Din.* 1.
49 Around twenty of the eighty-two biographical fragments concern the subject's death. See Bollansée (1999) 141–53, 179.
50 See, e.g., Regenbogen (1950) cols. 1418–53; Blum (1991) 124–247.
51 E.g., Regenbogen (1950) cols. 1428–29, using Ps.-Plutarch's *Vitae decem oratorum* as an example.
52 Cf. Dionysius of Halicarnassus, *Din.* 10–13.
53 See Lefkowitz (2012).

life of Horace notoriously notes that its subject's father was either a freed auctioneer's clerk, or a salt-fish merchant accused of wiping his nose on his sleeve. The first detail relies on confounding the persona assumed by Horace in *Satires* 1.6 with the poet himself, while Bion of Borysthenes, one of Horace's satirical models, claimed that his own father was of freed status and likewise wiped his nose with his sleeve.[54] If an author's biography can be (re)constructed from his books, then bibliography can serve as a potted biography or character witness. Thus Nepos writes of Atticus' books as though these are an extension of his identity:

> moris etiam maiorum summus imitator fuit antiquitatisque amator, quam adeo diligenter habuit cognitam, ut eam totam in eo uolumine exposuerit, quo magistratus ordinauit. [2] nulla enim lex neque pax neque bellum neque res illustris est populi Romani, quae non in eo suo tempore sit notata, et, quod difficillimum fuit, sic familiarum originem subtexuit, ut ex eo clarorum uirorum propagines possimus cognoscere. [3] fecit hoc idem separatim in aliis libris, ut M. Bruti rogatu Iuniam familiam a stirpe ad hanc aetatem ordine enumerauerit, notans, qui a quo ortus quos honores quibusque temporibus cepisset: [4] pari modo Marcelli Claudii de Marcellorum, Scipionis Cornelii et Fabii Maximi Fabiorum et Aemiliorum. quibus libris nihil potest esse dulcius iis, qui aliquam cupiditatem habent notitiae clarorum uirorum. [5] attigit poeticen quoque, credimus, ne eius expers esset suauitatis. namque uersibus, qui honore rerumque gestarum amplitudine ceteros Romani populi praestiterunt, exposuit ita, [6] ut sub singulorum imaginibus facta magistratusque eorum non amplius quaternis quinisque uersibus descripserit: quod uix credendum sit tantas res tam breuiter potuisse declarari. est etiam unus liber Graece confectus, de consulatu Ciceronis.
>
> He was the chief devotee of ancestral custom and a lover of antiquity, which he so diligently examined, that he explained it all in that volume in which he set the magistrates in order. For there is no law nor peace nor war nor illustrious deed of the Roman people, which is not noted under its appropriate date, and—which was very difficult—he wove beneath this the origin of the leading families, so that we can learn the descendants of famous men. He did the same independently in other books, such that he enumerated the Junian family from its source to the present, at the request of M. Brutus, noting who was born from whom, what offices he had held and at what times: he did the same for the Marcelli on behalf of Claudius Marcellus, and the Fabii and Aemilii for Cornelius Scipio and Fabius Maximus. Nothing could be more charming that these books for those who have some desire for familiarity with famous men. He also dabbled in poetry, so that he would not be lacking, I believe, in its sweetness. For it was in verse he expounded those who had surpassed the rest of the Roman people in honours and the magnitude of their deeds, such that he described, under the images of each of them, their deeds and magistracies in no more than four or five verses. It is scarcely to be believed that affairs of such moment could be set out so briefly. There is one book written in Greek, on the consulship of Cicero. (Nepos, *Att.* 18)

Charming, obliging to his friends, Atticus is above all one who wears his immense learning lightly and puts it at the disposal of others.[55] He dabbles in poetry so that he might not be without its 'sweetness'. *Suauitas*, in fact, is something of a motif of his biography. As a boy, it was the sweetness of his expression and tone

54 On these examples, see Power (2016).
55 On the thematic importance of friendship in the biography, see Stem (2005).

of voice that first captivated his lifelong friend Cicero (Nepos, *Att.* 1.3). The significance of this section of the biography should not be underestimated:[56] it is on the strength of this bibliographical sketch alone that Nepos, in assembling a biographical collection (the *De uiris illustribus*) that included volumes on Greek and Roman generals and Greek and Latin orators and historians, justified the inclusion of Atticus in the volume devoted to the latter.[57] Atticus' written works and historical interests are not otherwise mentioned in the remaining twenty-one sections of the life.[58]

Nepos' volumes on Roman orators and historians were clearly of use to Suetonius. Though no trace of a Suetonian biography of Atticus is preserved, he took from Nepos the detail that M'. Otacilius Pitholaus was the first freedman to write history (*Gram. et rhet.* 27.2), and cites him twice as an authority in the life of Terence (*V. Ter.* 2, 4). Though clearly quarried as a source of information by Suetonius, the wider literary-historical significance of Cornelius Nepos is harder to pin down; he seems to have been the first Roman author to write a chronicle (in the sense of a comparative chronology of history), the first to publish a collection of moral *exempla*, the first to write the biography of a living person and of a Roman knight (both achieved with the life of Atticus), and quite possibly the first to write political biography in general.[59] It may be the case that Nepos was also the first to write bibliography in this format (at least at Rome), though such an enquiry cannot be pursued further thanks to the fragmentary nature of our sources.[60] What is clear is that, of the models noted by Jerome, Suetonius' bibliographical practice most closely resembles that of Nepos.

---

56 See Horsfall (1989) 8: did Atticus himself request to be remembered in this way?

57 The manuscripts introduce Atticus' biography as an *excerptum e libro Cornelii Nepotis de Latinis historicis.*

58 We also possess an abbreviated life of Cato the Censor from this volume. Nepos had already written a monograph on Cato and preferred to summarise this for the *De uiris illustribus* (Nepos, *Cat.* 3.5). Cato's literary efforts take up a third of the brief sketch, but this again presents the image of the author as an historian, with an emphasis on *industria* (3.1, 4) and a detailed summary of the *Origines* (3.3–4). That Cato was a skilled farmer, lawyer, general, orator, and wrote speeches is also noted (3.1, 3); the possibility of alternative identities and biographies is thus acknowledged, but not explored (the occupations 'farmer' and 'general' even correspond to Cato's works *De agricultura* and *De re militari*, but these titles are not allowed to clutter the abbreviated life of the 'historian').

59 On Nepos' originality, see Geiger (1985); Horsfall (1989) xvi–xxi, 7–11; Tuplin (2000).

60 See Bardon (1952) 270–73. It is unclear whether Tiro's biography of Cicero dealt with bibliography: the fragments (Cornell *et al.* [2013] no. 46 F1–4) are all political. Political biographies by Oppius and Bibulus (Cornell *et al.* [2013] no. 40 F1–8, 49 F1) are even more unpromising, while M'. Otacilius Pitholaus' biography of Pompey (Cornell *et al.* [2013] no. 29 T1) had nothing to record (see n.23). Vulcatius Sedigitus' *De poetis* was a work of literary criticism in verse, though may have contained lists of Plautus' plays (Gellius, *NA* 3.3.1) and almost certainly those of Terence (Suetonius, *V. Ter.* 3): Courtney (1993) 93–96. Another list of Plautus' works was possibly provided by Aurelius Opillus (Gellius, *NA* 3.3.1 'Aurelius'); if in his *Pinax* (Funaioli [1907] 87 T4 = Suetonius, *Gram. et rhet.* 6.3) this will also have been in verse (note the acrostic). Ateius Philologus also wrote a *Pinax* (Funaioli [1907] 137 F1), contents unknown.

De uiris illustribus: Practice

Before examining Suetonius' practice in detail, some notes are required on the survival of the individual lives ascribed to Suetonius' *De uiris illustribus*. The life of Terence that introduces the Donatan commentaries is explicitly attributed to Suetonius (*Ter.* 8: *haec Suetonius Tranquillus*). Material found in the unattributed life of Horace that opens some manuscripts of his works is explicitly credited to Suetonius by Porphyrio, *Hor. Epist.* 2.1 (cf. *V. Hor.* p. 46.9 – 12 Reifferscheid). No authorial attribution accompanies the life of Vergil from the opening of Donatus' commentary, but stylistic considerations demonstrate that this biography is also the work of Suetonius (as Donatus' reliance on Suetonius for the life of Terence might in any case have suggested): with the exception of sections 37 – 8 and probably 17 (the latter discussed below), the life can be assumed to be the unadulterated work of Suetonius. Not only does Donatus state in the surviving preface of his commentary that he will report his sources verbatim, adding only a very little of himself (Don. *Epist. L. Mutatio*, p. 5 Hardie), but nothing else in the language or content of the life is demonstrably later.[61] The promise made in the Vergilian commentaries seems to have applied to Donatus' work on Terence as well: posited interpolations do not withstand scrutiny,[62] and while some maintain that certain quotations have fallen out of the life (all belong to a single poet, Vulcatius, a coincidence that should automatically raise one's sus-

61 See Körtge (1901) 19 – 20, 205 – 206; Greer (1926); Bayer (2002). These and other early studies, having established the Suetonian authorship of the life, were more concerned with hunting for possible interpolations (see the list of proposals in Brugnoli/Stock [1997] xv–xviii). However, this search has been shown to be methodologically flawed, and the absence of Donatus' stylistic 'tells' (along with other considerations) is strong evidence that, with the exception of sections 37 – 8 (plus 17, unless a later interpolation?), he did not intervene in the life: see the summary in Naumann (1981) (with additional bibliography); Stok (2010).

62 Pointing to the large number of learned authorities cited in the life of Terence compared to those of Vergil or Horace, Paratore (1946) 165 – 66, suggests that Donatus has added numerous scholarly citations to the life (less drastically, that Donatus also replaced a couple of Suetonian pronouns with *Terentium:* 235 n. 201). However, I would point out that of the authorities and works named in the Suetonian life (Fenestella; Cornelius Nepos; Porcius Licinius; Vulcatius; Varro; C. Memmius, *Pro se*; Santra; Q. Cosconius; Afranius, *Compitalia*; Cicero, *Limo*; C. Caesar), only Varro, Memmius, Afranius, and Cicero are quoted in the Servian commentary tradition or 'Donatan' commentary on Terence (both ultimately dependant on Donatus), and that even these 'shared' authorities (with the likely exception of Varro) are not quoted via the same works. D'Anna (1956) dismissing Paratore's argument on statistical grounds, also notes that Suetonius would naturally cite more authorities for the contested facts of Terence's life than for the established facts of those of Horace or Vergil. The example provided by the so-called *Auctarium Donatianum* (*V. Ter.* 8 – 10) that follows Donatus' reference to Suetonius (containing additional notes and citations by Donatus himself) should also be taken into consideration: the very existence of this coda implies that Donatus did *not* work his own research into the life. See Terzaghi (1931) 22.

picions),[63] only one obvious lacuna disfigures the text (between *V. Ter.* 3 and 4; see below). The life of Horace is likewise very largely free from later disturbance: a lacuna proposed so that a catalogue of Horace's works can be inserted (at p. 47.16 Reifferscheid) is neither prompted by the paradosis nor supported by Suetonius' usual practice (see below).[64] As the surviving lives of Terence, Horace, and Vergil faithfully represent Suetonius' original texts (with the minor caveats noted above), these are used in the following discussion alongside the *De grammaticis et rhetoribus* as authentic witnesses to Suetonius' bibliographical practice.

Several much shorter literary lives are also ascribed to Suetonius, but these will not form part of the present study because of their lack of relevant information or the textual challenges they pose.[65] These include lives of Tibullus,[66] Persius,[67]

---

63 Vulcatius is named three times in the Suetonian life: *V. Ter. 3* (= Courtney [1993] F2, one line); *V. Ter.* 5 (= Courtney [1993] F4, three lines); *V. Ter.* 7 (a list of the five authors Vulcatius ranked above Terence, not in direct quotation). If we accept a likely emendation of *uallegius* to *Vulcatius*, three additional lines of his poetry (= Courtney [1993] F3) are cited in the *Auctarium Donatianum* (*V. Ter.* 9). The passage in which Vulcatius ranked the poets is actually preserved in Gellius, *NA* 15.24 (= Courtney [1993] F1); because Gellius and Donatus can cite more of this obscure poet (otherwise only known from Pliny, *Nat.* 11.244), it is assumed that their source must have been a fuller version of the life by Suetonius (in Donatus' case this material must be assumed to have reached him by some devious and anonymous route): Schmidt (2000) 39. But Gellius quotes more of the passage than Suetonius would have needed to make his point (Courtney [1993] F1.1–4), and the familiarity of Pliny with this poet (hailed as *inlustris in poeticam*) at least offers a second potential route (Pliny's *Dubius sermo*) via which his fragments could have reached the later grammarians.

64 This is the only textual disruption generally agreed upon: see Schmidt (2000) 37 (with bibliography). Schmidt additionally conjectures here that details of Horace's early life may have dropped out before *V. Hor.* p. 44.7 Reifferscheid, pointing to their presence in the *Vita* of Porphyrio. However, the only information the scholiast adds is drawn entirely from *Serm.* 2.1.34–5 and *Epist.* 2.2.41–8, which hardly suggests dependence on Suetonius.

65 Suetonius' *De poetis* has unsurprisingly served as a convenient dustbin into which to tidy away other odd biographical fragments, many relating to individuals not certainly included in the work, including Laevius, Calvus, Q. Mucius Scaevola, Cinna, Ticidas, Propertius, Valgius, Domitius Marsus, Lentulus Gaetulicus, Cornificius, Cornelius Gallus, and Varius. Only the last three, however, are named among Jerome's excerpts. For this list, see Schmidt (2000) 32–33, 36 (Varius). It must be remembered that Suetonius had no monopoly over biographical trivia.

66 The Suetonian origin of a very brief life of Tibullus (= F27 in the edition of Rostagni [1944]) that follows the *corpus Tibullianum* in manuscripts of that author is now generally accepted: Schmidt (2000) 38 (with bibliography). However, besides the suspicion that the life has been severely shortened, there are signs of structural disruption (the quote of Domitius Marsus that prefixes the life must have originally been found at its conclusion), and of later tampering with the bibliographical section: *hic multorum iudicio principem inter elegiographos obtinet locum. epistolae quoque eius amatoriae, quamquam breves, omnino utiles* (*subtiles* Baehrens) *sunt*. The note on the 'utility' of the *epistulae amatoriae* was clearly made by a later grammarian, and the identity (or even existence) of these *epistulae* is uncertain: does this note refer to a selection of poems in Book 3 (those dealing with the love of Sulpicia and Cerinthus: Rostagni [1935]); to some separate and now lost work (Salanitro [1938] 46–47); or is it simply a medieval misunderstanding (Paratore [1946] 253–54)?

Lucan,[68] Passienus Crispus,[69] Calpurnius Piso,[70] and Pliny the Elder. Only the life of Pliny is explicitly attributed to Suetonius (in the manuscript incipits: *Vita Plinii ex catalogo uirorum illustrium Tranquilli*),[71] and this biography seems to have been severely abridged in transmission: while the short life as currently preserved contains no obvious gaps, it is hard to imagine that the favoured uncle of Suetonius' patron, Pliny the Younger, should have received such cursory treatment; details expected of a Suetonian biography (such as the subject's age at death or year of birth) are also absent.[72] Later additions probably include the alternative account of Pliny's death at the hands of one of his slaves,[73] and certainly include the introductory blurb to the *Naturalis historia* that now concludes the life. The surviving text mentions only two works by Pliny (the *Naturalis historia* and twenty volumes on the German wars), though we know the titles of five more.[74] In view of the problems posed by the transmission of this biography, it forms no further part of this discussion.

We turn now to consider Suetonius' bibliographical practice.

Suetonius seems to have consciously avoided the lists associated with pinacographic bibliography: none are included in the surviving *De grammaticis et rhetoribus* or *Vitae Caesarum*. That being the case, the conjectural 'restoration' of two

---

67 The life of Persius that is attributed to a (pseudepigraphal) commentary by Valerius Probus in our manuscripts is an awkward amalgam of two separate sources (one of which is likely to be Suetonius: see e. g., Rostagni [1935] 167–68) and shows clear signs of structural confusion (e. g., notices regarding the place and date of Persius' death are widely separated; details concerning Persius' youth, provided after the second notice of his death at the very end of his life, must have originally followed the treatment of this topic towards the opening of the biography, etc.): Martina (1984) 166–89.

68 The anonymous biography found in the *commentaria Bernensia* (Rostagni [1944] F31b) lacks its opening and contains several large lacunae, but preserves a first-person observation concerning public readings of Lucan in the author's youth, and closely agrees in tone and detail with the Suetonian fragment preserved in Jerome *Chron.* a.A. 2079 = 63 CE (= Rostagni [1944] F31a). The identification of the author as Suetonius has been accepted since Scaliger. A second work (Rostagni [1944] 176–86) that contains much biographical material, written in the fifth century or later (Ahl [1971] 1 n.1) and attributed to a certain Vacca, also drew upon an excellent biographical source (generally agreed to be Suetonius, see Schmidt [2000] 37 [with bibliography]) and contains much material missing from the former life, including the titles of eleven lost works (Rostagni [1944] 185.63–186.67).

69 A fragment of his Suetonian life from the *De oratoribus* is preserved by Jerome (*Chron.* a.A. 2054 [= 38 CE]), which agrees with a short biography preserved as a scholion to Juvenal 4.81 (= F71 Reifferscheid). See Baldwin (1983) 440; Reeve (1984) 47–48; Jones (1986) 245–51. The opening line of a senatorial address is quoted in the biography as part of an anecdote, but no other works are mentioned.

70 A brief biography of Calpurnius Piso is preserved in a scholion to Juvenal 5.109 printed by Giorgio Valla in 1486 and attributed by him to a certain 'Probus'. In view of the Suetonian origin of the life of Passienus Crispus in the Juvenal commentary tradition, and certain stylistic considerations, Jones (1986) 245–51 has also identified this biography as the work of Suetonius. However, none of the literary accomplishments suggested by the *Laus Pisonis*, 25–71, 163–65 are mentioned in this life.

71 For the history of the text (= F80 Reifferscheid) see Reeve (2011).

72 Baldwin (1983) 401–402 (also noting that Pliny's career is dismissed summarily).

73 Baldwin (1983) 403–404.

74 See the bibliography in Pliny, *Epist.* 3.5.3–6.

lists of titles to the lives of Terence and Horace would appear to be unsound (see below). A list of juvenilia transmitted with the biography of Vergil that can also be challenged on external grounds is thus doubly insecure:

> poeticam puer adhuc auspicatus in Ballistam ludi magistrum ob infamiam latrociniorum coopertum lapidibus distichon fecit ... deinde *Catalepton* et *Priapea* et *Epigrammata* et *Diras*, item *Cirim* et *Culicem*, cum esset annorum XXVI. cuius materia talis est ...
>
> Entering on poetry while still a boy, he composed a distich on Ballista, a school master stoned to death because of his evil reputation for highway robbery: ... Then *Catalepton* and *Priapea* and *Epigrammata* and *Diras*, also *Ciris* and *Culex*, when he was twenty-six years old. The subject of the latter is this ... (Donatus [= Suetonius] *V. Verg.* 17)

The catalogue of spurious works which today are known as the *Appendix Vergiliana* is unlikely to have appeared in Suetonius' original biography for several reasons, not least because such a bare list of titles is the sort of information that a grammarian would be itching to supply for his pupils but is otherwise unparalleled in Suetonius' surviving works.[75] Firstly, these titles do not appear in the verse life of Vergil by Phocas, which either relied independently on Suetonius, or on an early version of Donatus.[76] Secondly, the list includes poems with a great deal of apparently autobiographical content that do not inform the rest of Suetonius' biography.[77] Thirdly, only the *Culex* and *Catalepton* are attested before Suetonius; the *Dirae*, *Epigrammata*, and *Ciris* are first attested here (which, it should be remembered, is Suetonius' life of Vergil as transmitted by the fourth-century grammarian Donatus).[78] Even if arguments for the dependence of the *Ciris* on Statius have failed to gain widespread acceptance,[79] the work is still post-Ovidian, and if the *Dirae*, *Epigrammata*, and *Ciris* were accepted as Vergilian by Suetonius and his contemporaries, why did they leave no mark on the intervening centuries? Fourthly, the concluding note *cum esset annorum XXVI* is clumsy and ambiguous, and suggests violence has been done to the preceding text. The contents of the Vergilian list are thus problematic, but even if the above arguments are not considered decisive enough to challenge its basic authenticity, a list of titles remains an anomaly among Suetonius' surviving works and requires some special explanation, one that unfortunately is not easy to

---

75 Bayer (2002) 177–81 finds Suetonian parallels for the format of the list, but the pattern *deinde ... (et ...) item ... (et ...)* is so common that this is hardly decisive (cf. e.g., Columella, 9.5.6; Pliny, *Nat.* 29.26).

76 Favouring Suetonius: Strzelecki (1936); Mazzarino (1973–74); Mazhuga (2003); cf. Schmidt (2000) 34–36. The majority view holds that the work derives from Donatus: see Brugnoli (1984) vi–viii, with earlier bibliography, and now McGill (2010) 153–54, with new arguments and more recent bibliography. Phocas / Focas most likely belongs to the late fourth / early fifth century: Kaster (1988) 339–41.

77 Munari (1944) 258.

78 Lyne (1971) 235–36.

79 Lyne (1971) 233–53; Lyne (1978) 48–64; cf. Clarke (1973) 119–21. Against, see e.g., Thomas (1979) 181–82.

supply: Suetonius presumably wished to dismiss the juvenilia in summary fashion, but why did he bother to mention it at all? Like the grammatical titles suppressed in the *De grammaticis*, the titles (and even contents) of the *Appendix Vergiliana* added nothing to his biographical portrait.

I now turn to consider book lists attributed to Suetonius by modern conjecture. Certain readers of the life of Horace have found, in a possible lacuna signalled by an abrupt change of subject, a convenient space for the complete list of Horace's works which, given their familiarity with authors such as Diogenes Laertius and Jerome, they otherwise feel to be lacking from a biography:

> uixit plurimum in secessu ruris sui Sabini aut Tiburtini, domusque eius ostenditur circa Tiburni luculum. ⟨* * *⟩ uenerunt in manus meas et elegi sub titulo eius et epistola prosa oratione quasi commendantis se Maecenati, sed utraque falsa puto; nam elegi uulgares, epistola etiam obscura, quo uitio minime tenebatur.
>
> He lived chiefly in the seclusion of his estate, Sabine or Tiburtine, and his house is pointed out near the little grove of Tiburnus. There have come into my hands both elegies under his heading and a letter in prose, as if recommending himself to Maecenas, but I think both are fake; for the elegies are common, and the letter is obscure, a fault to which he was not at all subject. (Suetonius, *V. Hor.* pp. 47.15–48.3 Reifferscheid)

Otto Jahn first suggested a catalogue of the genuine works of Horace should have preceded Suetonius' note on the two spurious items,[80] but the life itself supplies this information.[81] As for the supposedly abrupt change of subject between the house(s?) near Tibur[82]—one of which was still pointed out as the home of Horace—and the false elegies and letter that have come into Suetonius' hands, the connection is surely that the tourist trap and fake writings both attest to the continuing popularity of the poet. Might the context also suggest that Suetonius held the house identified by the local guides in some suspicion: was it too new, or perhaps ill-fitted to his conception of Horace's tastes?

A second bibliographical list has been (mis)attributed to the life of Terence.[83] Though no convenient gap for the insertion of such a list is offered by this work, the biography as preserved in Donatus' commentary has at least one lacuna and may have been shortened elsewhere.[84] The titles of five out of the six plays constituting Terence's canon are woven into the surviving biography; only the *Phormio* is missing. Given this biography's state of preservation, it cannot be ruled out that the *Phormio* was mentioned in some lost passage, in which case, the life itself would surely take the place of a list. However, as the examples discussed below demonstrate, Suetonius was not overly particular in the matter of bibliographical com-

---

80 Ap. Reifferscheid (1860) 390.

81 See below. Dismissed on similar grounds by Naumann (1979) 151–52.

82 Cf. Lyne (1995) 9.

83 D'Anna (1956) 36–41.

84 See above.

pleteness. The absence of the *Phormio* might equally be attributable, therefore, to the idiosyncrasies of Suetonius himself; in some mitigation, he does begin his discussion of the comedies by pointing out that Terence wrote the magic figure of six plays (*V. Ter.* 3). More will be said about the structure of Terence's life shortly.

To summarise my argument so far, there is no reason to suppose that Suetonius attached pinacographic or 'Callimachean' book lists to any of his Lives. Jerome, despite claiming to follow Suetonius as his chief model, gives a misleading impression of his template in this regard.[85] However, Jerome does provide an instructive comparison in another respect. Though he professes to have conducted a systematic biographical and bibliographical trawl through Christian literary history, his work was hardly disinterested. Consider his handling of Tertullian. Jerome begins Tertullian's life with a note on his origins and floruit, before adding: *multaque scripsit uolumina, quae quia nota sunt pluribus, praetermittimus* (Jerome, *vir. ill.* 53.2: 'and he wrote many volumes which we will pass over, because they are known to so many'). Having dismissed Tertullian's works as too famous to merit listing, we then find out the alarming detail that their author fell into heresy in his later years, at which point Jerome's bibliographical scruples are suspiciously rekindled:

> specialiter autem aduersum ecclesiam texuit uolumina, de pudicitia, de persecutione, de ieiuniis, de monogamia, de ecstasi libros sex, et septimum, quem aduersum Apollonium composuit. ferturque uixisse usque ad decrepitam aetatem, et multa quae non exstant opuscula condidisse.
>
> In particular, he constructed against the Church books on modesty, on persecution, on fasting, on monogamy, six books on ecstasy, and a seventh that he wrote against Apollonius. He is said to have lived into very old age, and to have composed many essays which are not extant. (Jerome, *vir. ill.* 53.4)

Jerome's intention is to foreground Tertullian the heretic by listing a selection of his supposedly heretical works, and, by suppressing the rest of his impressively large *oeuvre* (over forty works,[86] many of which were not doctrinally suspect), to minimise his prominence as the first major Christian Latin author.

While questions of Christian orthodoxy clearly never troubled Suetonius, he too has his axes to grind. Similar absences with similar tendentious goals, made for local, rhetorical ends can be found in Suetonius' literary lives. Perhaps the most egregious instance is that of Q. Remmius Palaemon, whom Suetonius treats in a remarkably vicious manner. We learn that he was a woman's home-bred slave, engaged extensively and successfully in trade (*Gram. et rhet.* 23.1, 6), and was extraordinarily

---

85 Cf. the lists of titles at Jerome, *vir. ill.* 1.5, 5.9, 11.4–5, 16.2, 23.2, 24.2, 25.2, 26.2, 27.2, 35.4, 38.3, 53.5, 61.2, 66.1, 69.5–6, 70.2, 74.2, 80.2, 81.2, 83.1, 87.2, 90, 91.2, 94, 100.3, 109.2, 116.1, 117.2, 3, 135.2–5. Jerome adopted this format because much of his early information was drawn from Eusebius (Jerome, *vir. ill.* pr.5), who frequently did resort to lists of books. See the parallels collected in the commentary of Halton (1999).

86 See Tränkle (2000).

arrogant, extravagant, and sexually depraved—a monstrous Trimalchio with a gift for teaching. The strength of the attack suggests that Suetonius relied on some lost contemporary invective, possibly by Seneca, who loathed Palaemon passionately (Pliny, *NH* 14.51).[87] What we do not learn from Suetonius' life, but what Suetonius must himself have known, is that Palaemon was the author of an extremely successful—even influential—grammatical handbook (or *Ars grammatica*): the contemporary fame of this work is attested by Juvenal, Pliny, and Quintilian.[88]

If Palaemon's titles have been suppressed to belittle the individual in question, the scholarly works of Staberius Eros, which including at least a treatise *De proportione*,[89] may have been deliberately omitted, as Barry Baldwin has suggested,[90] to amend the inflated claims regarding his importance made by Pliny the Elder. Pliny, with more enthusiasm than judgement, had characterised Eros as *grammaticae conditor* (*Nat.* 35.199); not only was his contribution to the discipline too insignificant to merit a mention in the preface of the *De grammaticis*, but Suetonius gave him one of the shortest biographies found in the work.

Now, Suetonius may have merely wished to avoid telling his audience what they already knew, particularly in the case of the famous Palaemon,[91] but the suppression of such titles also allows him to score cheap points. And besides, even unfamiliar and recherché material had its limits, as the case of Aurelius Opillus demonstrates:

> conposuitque uariae eruditionis aliquot uolumina ex quibus nouem unius corporis quae, quia scriptores ac poetas sub clientela Musarum iudicaret non absurde et fecisse et ⟨in⟩scripsisse se ait ex numero diuarum et appellatione. huius cognomen in plerisque indicibus et titulis per unam ⟨L⟩ litteram scriptum animaduerto, uerum ipse id per duas effert in parastichide libelli qui inscribitur Pinax.
>
> He wrote several books of varied scholarship, including nine as a single work which, he says, he composed and named after the number and name of the goddesses—since he considered (not unreasonably) that poets and authors are under the patronage of the Muses. I notice that his surname has been written with one L in many book-labels and titles, but in fact he himself spells it with two in the acrostic of the little book which is called Pinax. (Suetonius, *Gram. et rhet.* 6.2–3)

Unless the only books that Suetonius had checked were the *Musae* and *Pinax*, he knew more of Opillus' *uolumina* that he was prepared to report, including specific titles (since he carefully recorded the spellings of their author's name). If novelty

87 Kaster (1995) 231–32.

88 Juvenal, *Sat.* 6.451–3 (cf. 7.215–19); Pliny, *NH* 15.49; Quintilian, *Inst.* 1.4.20. See Kaster (1995) 230–31.

89 Funaioli (1907) 107 F[1].

90 Baldwin (1983) 433.

91 Cf. Lindsay (1995) 69–70. Though Lindsay claims the omission of commonplace material on a Horace or Vergil might be acceptable to Suetonius' audience (using the specific example of book titles), he does not take into account similar gaps found in many of the minor biographies of the *De uiris illustribus*.

was the sole criterion for bibliographical inclusion, the rest of Opillus' *uolumina*—apparently as deeply obscure to Suetonius' contemporaries as they are today—should also have merited mention.[92]

A contrasting example of Suetonius' bibliographical manipulation is offered by Cornelius Epicadus:

> Cornelius Epicadus, L. Corneli Sullae dictatoris libertus calatorque in sacerdotio augurali, filio quoque eius Fausto gratissimus fuit; quare numquam non utriusque se libertum edidit. (2) librum autem quem Sulla nouissimum de rebus suis imperfectum reliquerat ipse suppleuit.
>
> Cornelius Epicadus, freedman of the dictator Lucius Cornelius Sulla and his servant in the augural priesthood, was also most highly esteemed by his son, Faustus; thus, he always put it about that he was the freedman of both. In addition, he himself finished the book that Sulla had left last and unfinished concerning his actions. (Suetonius, *Gram. et rhet.* 12.1)

As the life of a notionally famous teacher of grammar, the biography of Cornelius Epicadus must be judged a complete failure. We are told nothing at all about his pedagogical or scholarly activities: the completion of the dictator Sulla's memoires was a work of homage that required the insight and sympathy of an old retainer, not grammatical expertise.[93] Suetonius could have added to this biography Epicadus' grammatical writings: his work *De metris* was likely the first such treatise written in Rome;[94] we also know of a *De cognominibus*.[95] Other fragments suggest Epicadus wrote works of an antiquarian character, though their titles are now lost.[96] Suetonius might be excused for his failure to mention the grammatical works if he did not know of their existence, but why, then, include Epicadus among the grammarians? On what else is his reputation likely to have been based in the second century CE?[97] He was presumably the live-in tutor of Sulla's son but (apparently) did not even open his own school. Suetonius may well have relied on the dedicatory preface of

---

92 Besides Gellius, *NA* 1.25.17 (= Funaioli [1907] 88 F2), the remaining witnesses are early or derive their knowledge from early sources. Funaioli (1907) 88–93 F3, 5, 11, 15, 18, 23 come from Varro. Funaioli (1907) 88–93 F4, 6–10, 12–14, 16, 19–22, 24 come from Festus' abridgement of Verrius Flaccus, Funaioli (1907) 91 F17, p. 93 F25 come from elsewhere in the grammatical tradition, but have similar early origins. F25 (= Char. 162.19–23B), a fragment of Julius Romanus, likely came from Pliny's *Dubius sermo* (= p. 49.17 Beck); as Opillus is not otherwise mentioned in the fragments of this work or in Pliny *NH*, one suspects the citation came from an intermediary (Varro?). For a passage in Fronto (p. 15.15 vdH) that refers to an MS of Titius copied by an Aurelius (the name is corrupt), see Kaster (1995) 111. Clearly, no one but Suetonius was actually reading this author *ca.* 100 CE.

93 Cf. the similar language used to describe Hirtius' continuation of Caesar's memoires: Suetonius, *Iul.* 56.1.

94 Funaioli (1907) 104 F2: see Kaster (1995) 55–56.

95 Funaioli (1907) 104 F1.

96 Schanz/Hosius (1927) 581.

97 Epicadus is otherwise known as a name to Arnobius (*Adv. nat.* 1.59) and Macrobius (*Sat.* 1.11.47), but is cited via Varro at Servius Auctus, *Aen.* 1.649.

one of his grammatical works—doubtless addressed to Faustus Sulla—for the meagre biographical details he has collected above.

Clearly, Suetonius is interested in Epicadus only because he shared a connection with Sulla and his son. This short biography has, in fact, been carefully shaped to fit into a much larger pattern, obvious only when the *De grammaticis* is read in a body: Suetonius is not concerned with the intellectual discipline of grammar or the development of teaching, but instead with human stories and the growth of the profession's independence.[98] Beginning with early practitioners who were beholden to the households of the Republican dynasts, Suetonius records the gradual rise of the self-supporting teacher who, like Verrius Flaccus, could aspire to his own honorific statue (*Gram. et rhet.* 17.3) or, like Remmius Palaemon, could earn a yearly teaching income of HS 400,000 (23.5). In the specific case of Epicadus, the colourless books on grammar added nothing to his image as the loyal servant of the Cornelii, and were thus superfluous.

Verrius Flaccus provides yet another case. If Epicadus is an example of the humble status enjoyed by early grammarians, and Remmius Palaemon represents the profession in its later pride and arrogance, then Verrius Flaccus occupies a pinnacle of tasteful excellence. He was invited by Augustus to teach his grandchildren (with the stipulation that he accept no new students thereafter), and accordingly moved his school to the Palatine (17.2). He is one of two grammarians named by Suetonius to have been honoured with statues;[99] that of Flaccus stood in the forum of Praeneste, near his monumental *Fasti* (17.3); fragments of this calendar still survive. This is the only scholarly work, however, mentioned in the biography by Suetonius, who prefers to dwell on Flaccus' excellence as a teacher, noting his use of old books as prizes for star pupils (17.1). The roll-call of his learned titles, including *Res Etruscae*, *De obscuris Catonis*, *Saturnus*, and the colossal lexicon *De uerborum significatu* (the P-words alone filled at least five volumes),[100] seems to have been considered a distraction from the image of an inspirational instructor. As the learned works of Epicadus added nothing to the portrait of the faithful retainer of the Cornelii, so those of Flaccus make no contribution to the retainer of Augustus: at the opening of his life, Suetonius is keen to stress that *docendi genere maxime inclaruit* (17.1: 'he grew famous chiefly from the style of his teaching'). It was for this reason (and implicitly *not* because of his written works) that Augustus appointed him as a tutor to his grandsons (17.2). Note, finally, the casual way in which Suetonius signals the existence of Flaccus' *De orthographia* in the subsequent life of Scribonius Aphrodisias (19.2): a marker, perhaps, that silence should not be confused with ignorance.[101]

---

98 Kaster (1995) xliii–xlvi.

99 The other is L. Orbillius Pupillus (*Gram. et. rhet.* 9.6).

100 Schanz/Hosius (1935) 361–67.

101 Baldwin (1983) 438–39, suggests that Suetonius' silence on Verrius Flaccus' works ('not even a vague *scripsit multa*') may be connected with the critical reception he receives from Aulus Gellius, but that author, though ready to register his dissent from Flaccus when merited, is hardly contemptuous.

A number of other grammarians are credited with the writing of 'numerous works', 'several pamphlets', 'a number of grammatical books' etc., with only one or two titles selected by name for any comment:

Table 1

Suet. *Gram. et rhet.*	**Titles known from elsewhere**[102]
5. Servius Nicanor: commentarii; satura.	
6. Aurelius Opillus: uolumina uariae eruditionis; Musarum libri IX; Pinax.	
7. Antonius Gnipho: uolumina De Latino sermone II; (Philologus tradit cetera scripta discipulorum eius esse).	commentarius in Ennium.[103]
8. Pompilius Andronicus: multa composuit; opusculum Annalium Enni elenchorum.	
9. Orbilius Pupillus: quiddam scriptum; Περὶ ἀλογίας.	*De differentiis uerborum?[104]
10. Ateius Philologus: commentarii (quamquam paucissimi extent); epistula ad Laelium Hermam; Ὑλῶν libri DCCC; breuiarium rerum omnium Romanarum pro Sallustio; praecepta de ratione scribendi pro Asinio Pollione.	liber inscriptum 'an amauerit Didun Aeneas'; Πίνακες; Glossemata.[105]
11. Valerius Cato: Indignatio (libellus); grammatici libelli; poemata: Lydia, Diana.	
12. Cornelius Epicadus: commentarii rerum gestarum Sullae.	De cognominibus; De metris; *Antiquitates?[106]
13. Staberius Eros: –	De proportione.[107]
14. Curtius Nicias: de Lucilio libri.	
15. Pompeius Lenaeus: satura.	*De materia medica.[108]
16. Caecilius Epirota: –	
17. Verrius Flaccus: fasti Praenestini; De orthographia libri (*Gram. et rhet.* 19.2).	poemata; Res memoriae dignae; Res Etruscae; Saturnus; epistulae; De obscuris Catonis; De uerborum significatu.[109]
18. Crassicius Pasicles (Pansa): Commentarius Zmyrnae.[110]	
19. Scribonius Aphrodisius: contra libros Verrii De orthographia.	
20. Iulius Hyginus: [De uita inlustrium uirorum (? = De uiris claris). Cf. Jerome, *vir. ill.* pr.].	

See Holford-Strevens (2003) 162. Flaccus' epitomator Festus, working around the same time as Gellius, evidently had no major qualms about the 'respectability' of his source.

102 See generally Kaster (1995) ad locc.

103 Funaioli (1907) 99 F1.

104 Cf. Funaioli (1907) 135 F1–2. (* = title restored by conjecture; ? = existence of work speculative).

105 Funaioli (1907) 137–8 T9, F1–2.

106 See above.

107 See above.

108 Cf. Funaioli (1907) 404 T5.

109 Funaioli (1907) 510–14 T7 (= Courtney [1993] 291) T8–10, F1–4.

110 Though see Kaster (1995) 197, for references to a Pansa who may or may not be Crassicius.

Table 1 *(Continued)*

Suet. *Gram. et rhet.*	Titles known from elsewhere[102]
	De agri cultura (? = De apibus); commentarius in Cinnam; libri de Vergilio; [De uita rebusque inlustrium uirorum (? = De uiris claris)]; Exempla; De familiis Troianis; *De origine et situ urbium Italicarum; De proprietatibus deorum; De dis penatibus.[111]
21. (Maecenas) Melissus: Ineptiarum / Iocorum libri CL; opera diuersa; fabulae trabeatae.	De apibus (? = *Naturalis historia); *De dubiis nominibus?; *De proprietate sermonis?[112]
22. Pomponius Porcellus: –	
23. Remmius Palaemon: poemata.	Ars grammatica.[113]
24. Valerius Probus: (editiones); pauca de quibusdam minutis quaestiunculis.	epistula ad Marcellum; commentarii; commentarius de occulta litterarum; De inaequalitate consuetudinis; De temporum conexione; De genetiuo Graeco; notae.[114]

Looking at this list, it is apparent that the works which generally interest Suetonius are the more literary or unusual items, scraps of poetry or clever titles.

Above, I proposed that Suetonius was prepared to suppress mention of written works in a biography for local rhetorical goals. Collected references of the 'many books' and 'several *libelli*' suggest that Suetonius perceived a wider, structural problem with his chosen approach to biblio-biography. Though the use of summary categories of the sort encountered in Diogenes Laertius' biography of Timon (see above) might be presumed to depend on the abridgement of some pre-existing pinacographic bibliography in the Greek biographical tradition, this assumption is inapplicable to large swathes of Roman biography.[115] Suetonius clearly felt that lists of titles were inelegant and should be avoided. If lives were to be written and woven around an author's written works, the mention of too many would distract from the often meagre biographical details that could be assembled. Rather than write biblio-biography, Suetonius was in danger of boring the reader with fat Callimachean bibliographies attached to thin Hermippan biographies, producing a work with a few juicy titbits disfigured by a succession of technical titles on obscure questions of grammatical trivia. Whatever the reason for the inclusion of Hermippus in Suetonius' preface (as reported by Jerome), it was clearly not for the model he provided as a bibliographer.

111 Schanz/Hosius (1935) 368–72; Funaioli (1907) 527–36 F1–19; Cornell *et al.* (2013) no. 63 F1–15.
112 Schanz/Hosius (1935) 176–77; Funaioli (1907) 538–40 T7, F4–6.
113 Mazzarino (1955) F1–8.
114 Schanz/Hosius (1935) 734–37 (apocrypha: Kaster [1995] 247–48).
115 E. g., Nepos wrote the life of a living subject (Atticus), while Suetonius pioneered the study of the grammarians (on his lack of sources, see Kaster [1995] xxvi–xxx).

Robert Kaster has suggested that Suetonius barely glances at his subject's written works because he is more intent on conveying a sense of their teaching activities.[116] One need only consider the case of Flaccus to see the truth of this statement. But lives like those of Opillus (or Caesar, Nero, Horace, or Vergil) suggest that Suetonius himself delighted in the technical aspects of bibliography.[117] The *De grammaticis* must have been the most challenging section of Suetonius' project, because no one had troubled to record the biographies of such marginal literary figures at Rome before.[118] But choosing to write his lives in and around his authors' bibliographies placed additional constraints upon Suetonius. To balance the pictures of men whose biographies were exceedingly thin, and whose written works on technical subjects gave little opportunity to write in the autobiographical mode, the bibliographical component of Suetonian biblio-biography had to be carefully managed.

The lives of the famous Roman authors of more prestigious literature—actual poets, real orators—had already attracted scholarly attention (thus Jerome can name Varro, Santra, Nepos, and Hyginus). These individuals gave Suetonius scope to write more rounded biblio-biographies. We have examined the lives of Vergil, Horace, and Terence to establish the absence of bibliographical lists in these works. But how, then, did Suetonius incorporate bibliography into the structure of these more ambitious lives? I will consider the lives of the three authors just mentioned, as these represent the most complex bibliographies surviving from the *De uiris illustribus.*

The lives of Terence, Horace, and Vergil each begin with the name, status, background, and (usually) birth of the author.[119] Then follows a central section that discusses their works,[120] with a conclusion concerning e.g., physical characteristics, death, and subsequent reputation.[121] Exactly the same structure is found in the biographies of the *Vitae Caesarum.*

Suetonius opens his discussion of Terence's comedies (3–4) by noting the playwright wrote six in total, and that the first of these, the *Andria*, caused a sensation on its first reading. The rest of the plays were similarly well received, though a certain Vulcatius (or Volcacius) has something negative to say about the *Hecyra* (the quotation is corrupt).[122] This was Terence's second play, though Suetonius does not mention this fact. Now we jump to the *Eunuchus* and *Adelphoe*, the fourth and final plays.

---

116 Kaster (1995) 231.

117 Note also his interest in library history and sophisticated use of codicological vocabulary Gomoll (1935) 381–88; Loyd (1969).

118 Kaster (1995) 231.

119 Suetonius, *V. Ter.* 1–2; *V. Verg.* 1–16; *V. Hor.* pp. 44.3–46.2 Reifferscheid (1860). The date of Horace's birth is delayed so that it can be reported alongside his death: p. 48.3–4 Reifferscheid (1860).

120 Suetonius, *V. Ter.* 3–4; *V. Verg.* 17–34; *V. Hor.* pp. 46.2–47.4 (apocrypha: pp. 47.17–48.3) Reifferscheid (1860).

121 Suetonius, *V. Ter.* 5–7; *V. Verg.* 35–46; *V. Hor.* pp. 47.5–48.9 Reifferscheid (1860).

122 See Courtney (1993) 88, 94–95.

The former is mentioned because it so pleased the audience that it was performed twice on the same day, the latter because Varro thought more of its opening than Menander's original (a lacuna seems to intervene between the two notices, though the missing material will relate to the public reception of the *Adelphoe*).[123] Opening the next section (4), the *Adelphoe* is cited for Terence's supposedly half-hearted defence against critics who claimed that his comedies were written with the help of his aristocratic patrons, Scipio and Laelius. This question of authenticity dominates the remaining discussion. The *Heautontimoroumenos*, Terence's third play, is mentioned last and only because Laelius—so said Nepos—revealed his authorship of certain lines to his wife. The *Phormio*, as previously noted, is omitted.

Table 2

Order of Terence's plays in Suetonius	Chronological Order[124]
Andria	*Andria*
Hecyra	*Hecyra* (first unsuccessful performance)
Eunuchus	*Heautontimoroumenos*
Adelphoe	*Eunuchus*
Heautontimoroumenos	*Phormio*
	Adelphoe

On recounting the death of the author, Suetonius tantalises the reader by reporting the loss of an additional corpus of plays: returning home after visiting Greece for inspiration, Terence was either lost at sea with a cargo of new works adapted from Menander, or watched as his new plays sank in the harbour and pined to death over their loss (5).[125] The motif of a failed homecoming with devastating bibliographical consequences is also encountered in Vergil's biography.

From this overview, we can see that Suetonius discusses Terence's corpus not in chronological sequence (nor that in which the plays appear in our manuscripts),[126] but according to his own sensationalising interests, first concentrating on Terence's runaway success, then undercutting this with doubts regarding whether a low-born African could have authored the comedies unaided.[127]

In Horace's biography, Suetonius begins his bibliographical section (*V. Hor.* pp. 46.2–47.4 Reiff.) *in medias res*. Augustus is already pleased enough by Horace's

123 Cf. Terence, *Ad.* didasc.

124 Lefèvre (2014) 247–48.

125 These tales may have been transplanted from the biography of Menander, who is also said to have drowned: Goldschmidt (2015), accessed 5 September 2017.

126 A (Cod. Bembinus): *An., Eun., Haut., Phorm., Hec., Ad.*; Γ family: *An., Eun., Haut., Ad., Hec., Phorm.*; Δ family: *An. Ad. Eun. Phorm. Haut. Hec.* Reeve (1984) 415–16.

127 The prominence given to the rebuttal of contemporary accusations of plagiarism in the plays themselves (Terence, *Ad.* 15–21; *Haut.* 22–4) doubtless explains the fixation of the subsequent biographical tradition on questions of authorship.

poetry that he invites him to compose the *Carmen Saeculare*, and to follow this up with a fourth book of *Odes* honouring Tiberius and Drusus. Suetonius adds that the previous three volumes had long been put aside by their author. The *Carmen Saeculare* was published in 17 BCE, thus the works pleasing to Augustus (in addition to *Odes* I–III) must have been the *Epodes* (not otherwise mentioned) and *Sermones/Saturae* I–II (all published by 30 BCE), and *Epistulae* I (completed *ca.* 20/19 BCE).[128] Other works termed *sermones* are next noted (p. 46.7): Suetonius says that after reading some of these, Augustus was disappointed to find he had not been mentioned and wrote a letter of rebuke to Horace. As quoted, the actual letter has a subtly different complaint, namely that Horace does not converse with him like he does others 'in your numerous works of this kind' (*plerisque eius modi scriptis:* p. 46.9 – 10). Suetonius states that Horace accordingly shot off the *ecloga* (used here in the sense of 'short poem') that begins '*cum tot sustineas* ...' (quoting the first four lines of *Epist.* 2.1, composed *ca.* 15 – 12 BCE). Considering the context in which Augustus' letter is said to have been written, critics have been quick to suggest that Augustus was responding to works that had appeared shortly before *Epistulae* II, i.e *Epistulae* I.[129] Whether Suetonius and Augustus are describing the same works, however, is a moot point. Horace uses the term *sermones* of both his satires and epistles, but given that *Sermones* was also used as the title for the former, Suetonius may well have only these works in mind.[130]

Suetonius then moves from Horace's works to his personal appearance (p. 47.5 – 12), but finds here an opportunity to mention that Horace himself admitted to being fat in his satirical medleys (a reference to the first book of *Epistulae*). Augustus apparently used the receipt of this work to call for yet greater efforts from Horace, so that the circumference of his next *uolumen* might match his belly (see below).

Suetonius' claims regarding Augustus' interventions in Horace's corpus have proved controversial.[131] What matters for the present discussion, however, is that once again the presentation of Horace's bibliography is entirely shaped by the thematic concerns of Suetonius and in particular, his interest in patronage. The treatment see-saws between earlier works that prompt a response from Augustus, and those that respond to his requests. Horace's *oeuvre* is thus represented as the product of negotiation with his imperial patron.[132]

---

128 On Horace's chronology, see e. g., Nisbet (2007) 7 – 21.

129 E. g., Fraenkel (1957) 383.

130 See Brink (1963) 241 – 42. Discounting the possibility that Augustus was reacting to a first reading, he may be making a simple request for a genre piece *à la Sermones* I–II. Cf. Rudd (1989) 1.

131 See e. g., Griffin (1984); White (1993) 112 – 55.

132 The complex use of Augustan patronage to supply the 'structure' of the bibliography entirely rules out the suggestion of Schmidt (1994) 492, that the bibliography was originally arranged in two generic sequences (lyric and hexameter poems). Any re-writing would have had the ironic effect of making the bibliography appear *more* Suetonian.

There is one glaring omission: the *Ars poetica*, known as such by Quintilian and later authors,[133] and included in Pseudo-Acro's Horatian catalogue.[134] The verse epistle is treated in modern editions as an appendix to *Epistulae* II for (supposed) chronological convenience, but the manuscripts generally place it second after the *Odes*, or fourth after the *Odes*, *Epodes*, and *Carmen Saeculare*.[135] The date of the *Ars poetica* is controversial.[136] Publication may have come before the *Carmen saeculare* (*c.* 19/18) or after *Odes* IV (*c.* 10 – 8 BCE), which leaves three possibilities. 1) The poem is implicitly included by Suetonius among the writings initially pleasing to Augustus referred to at p. 46.2. 2) The poem was thought by Suetonius to precede the *Carmen Saeculare* and postdate *Epistulae* I, in which case its existence might be obliquely acknowledged through the quotation of Augustus' wish (in response to *Epistulae* I?) that for his next project, Horace should fill up a 'swelling' (ὀγκωδέστατος) book-roll: the *Ars poetica* was by far Horace's longest poem, at 476 lines.[137] 3) The poem was thought to be Horace's last work, and any mention of it was deliberately suppressed by Suetonius. If the work is not ignored because it is lumped together with Horace's early poems (scenario no. 1 above), the dedication of the *Ars poetica* to the Cornelii Pisones (*Ars* 6) clearly could not be neatly fitted into the image of Augustan intrusion that dominates Suetonius' presentation of Horace's later works, and was perhaps overlooked for the sake of maintaining an Augustan focus (the poem also has little to offer the biographer, which suggests that Suetonius' interest in the work would, in any case, have been marginal at best).

Table 3

Order of Horace's poems in Suetonius	Chronological Order
(*scripta eius* = *Odes* I–III, *Epodes*, + *Ars poetica?*)	*Saturae* I
Carmen Saeculare	*Saturae* II; *Epodes*
Carmina (= *Odes*) IV	*Odes* I–III
Sermones (= *Saturae*) (+ *Epistulae* I?)	*Epistulae* I
[*Epist.* 2.1.1 – 4]	*Ars Poetica?*
Satura (= *Epistulae* I)	*Carmen Saeculare*
	Epistulae II; *Odes* IV
	Ars Poetica?

133 Quintilian, *Inst.* 8.3.60; Symmachus, *Epist.* 1.4; Sidonius Apollinaris, *Carm.* 22, 9.220; Priscian, 3 p. 254.16 Keil.
134 Ps.-Acro *V. Hor.* p. 3.6 Keller.
135 Tarrant (1984) 184.
136 See Brink (1963) 239 – 43; Syme (1986) 379 – 81.
137 While Suetonius' biography may not help to establish the absolute date of the poem, it may indirectly contradict Porphyrio (ad *Ars* 1), who identifies the addressee as L. Cornelius Piso, cos. 15 BCE (thus suggesting the later date).

The letter of Augustus and description of Horace's physical characteristics is followed by a notorious passage concerning Horace's sexual mores. But before Horace's death and the conclusion of the biography, Suetonius takes a moment to dismiss the two spurious works already encountered, one of them posing as a piece of juvenilia (the letter of recommendation to Maecenas). Pseudepigrapha thus form a coda to Horace's biblio-biography, whereas in the life of Vergil (see below), the *Aetna*, likewise of disputed authenticity and also purporting to be a work of the author's youth, is noted at the chronologically appropriate point (*V. Verg.* 19). Questions of authenticity are part of the learned Hellenistic baggage of bibliography,[138] and for good reason, as the chapter of Matthew Nicholls in this volume demonstrates. Although Suetonius' engagement with specific problems cannot be entered into here, it is important to note that such debates can be as much about establishing a rhetoric of *auctoritas* as actually getting to grips with cases of literary fraud. In Suetonius' case, the emphasis on his own judgement (*falsa puto, nam* ...) is again a means of using bibliography for rhetorical ends—here, establishing his own credentials as an expert. This is hardly a crushing contribution to one of those heated controversies that raged around the canon of Plautus (or, apparently, Nero):[139] tellingly, Suetonius gives no impression that anyone else knows of the existence of this material. Some minor pieces of misattributed ephemera have floated into his lap, but the opportunity for point-scoring was too good to miss.

We can now return to the life of Vergil, the final example to be examined in this study. We have already seen that the catalogue of spurious works found in his biography is probably a later addition. In his treatment of Vergil's authentic works, Suetonius now adopts a chronological approach. But his narrative is not a simple one of progress towards the goal of the *Aeneid*. Vergil, preternaturally talented, wrote an epitaph for Ballista, a teacher-turned-bandit, and the *Culex* (*V. Verg.* 17–18). He also possibly authored an *Aetna* (19). The inclusion of this work within the chronological sequence seems to strengthen its claims to authenticity (compare the treatment of the Horatian pseudepigrapha discussed above), but as the surviving *Aetna* was potentially forged in Suetonius' own youth, the passage itself may be open to doubt.[140] Vergil then attempted *Res Romanae* (i.e., the subjects that will eventually lead to the *Aeneid*), but found himself unequal to the task and turned aside to write the *Bucolics* and *Georgics* (19–20). The *Aeneid* follows (21). The *Georgics* were perfected by Vergil's editing process, which distilled a great deal of material into a very few lines and licked these into shape (22). The *Aeneid*, by contrast, is fundamentally flawed (23–24). In his hurry to write, Vergil left numerous line fillers and died before completing the final edit, initially requesting that the work be burned on his death-

138 See Speyer (1971) 112–28; Peirano (2012) 37–54.

139 Cf. Gellius, *NA* 3.3.3; Suetonius, *Nero* 52.

140 See Goodyear (1965) 56–59; Goodyear (1984) 358–59; Volk (2005) 69–72. However, note Servius Auctus, *Aen.* 3.571, summarising a Vergilian *Aetna* unlike the surviving version, hints at the existence of another now-lost poem; was *this* the work Suetonius doubted? See Goodyear (1984) 358–59.

bed. Suetonius claims that Augustus stepped in to compel Vergil's literary executor, Varius, to publish the work; the notorious half-lines were left uncorrected (39–41). Certain persons reported that Varius had changed the order of the books, and had deleted the original opening (42). Suetonius' claims have launched a great deal of scholarly debate.[141]

Once again, the structure of the bibliography is made to do more interesting work than simply track the growth of an author's corpus. In Vergil's case, the *Aeneid* is found in final position as a bibliographical anti-climax; however great it may be as a work of literature, without Augustus' intervention it would never have been published in its imperfect state. The *Aeneid* almost became a work that Suetonius could ignore with a clear conscience.

Each of the three lives investigated take a different approach to discussing and structuring bibliography. Rather than adopt the Alexandrian technique of arranging works by genre, Suetonius uses bibliography as a skeleton to support the flesh of his lives: in the case of Terence and Horace, their works can be used to explore themes of patronage and authenticity; in that of Vergil, composition and publication. There is a bibliographical link too between the Lives of Vergil and Terence: the former died embarking on a journey to Greece to finish the editing of the *Aeneid* and lost control of his work's final publication; the latter died supposedly returning from a journey to Greece, bringing back new plays that were lost with his ship. Thus neither quite fulfilled their bibliographical promise.

Conclusion

Suetonius' bibliographies are a tale of books ignored, books lost, and even books that were never meant to be. Why Suetonius wrote bibliography in this manner, eschewing the sober catalogues of the Callimachean tradition for embedded, incomplete, and gossipy discussion of selected titles, is surely to be answered, in part, by pointing to the urge that drove Hermippus to spice up his master Callimachus' *Pinakes* with racy biographical sketches. Suetonius' lives are enjoyable reading. Even those of the grammarians, a collection that might have descended into a tedious *Handbuch der grammatischen Literatur*, make for compelling entertainment. Suetonius still performed a scholarly service by writing about these individuals alongside the grander poets and historians; as previously noted, his was the first systematic treatment of such figures at Rome. Many of the names listed would be unknown to us but for Suetonius' biblio-biographies.[142] But his work is emphatically

141 See e.g., O'Hara (2010).

142 Otherwise unattested: Servius Nicanor; Pompilius Andronicus; Caecilius Epirota; Crassicius Pasicles; Scribonius Aphrodisias.

not the last word.[143] Where Callimachean or pinacographic bibliographies act as authoritative points of reference, establishing the boundaries of an author's corpus and leading users to their books, Suetonian biblio-biography entices the reader to dip in and out of literature, just as biographical collections encourage the reader to slip in and out of individual lives.

A second reason Suetonius avoided objective and systematic bibliography might be attributed to the institutional constraints encountered in Rome itself. Callimachus' *Pinakes* could only have been written in a universal library; any Roman imitator would have to master the holdings of several public institutions. Whether or not Suetonius composed his works while serving as *a bibliothecis*, in the 120s CE this office was notionally in charge of several physically distinct institutions. These included libraries in the Atrium libertatis, the Portico of Octavia, the Domus Tiberiana (possibly identical with an institution said to have been founded in the New Temple of Augustus), the Templum Pacis of Vespasian, a library next to the Temple of Apollo on the Palatine, and Trajan's pair of new libraries behind the Basilica Ulpia.[144] Though individual libraries may have held catalogues of parts (or even all) of their holdings,[145] institutional fragmentation militated against the creation of a comprehensive Roman national bibliography.[146] Into this space, however, Suetonius was able to step with his own idiosyncratic collection of lives, using bibliography in a much more flexible, and ultimately enjoyable, fashion.

---

143 Thus arguments about an author's corpus that depend upon Suetonius' silence are futile; cf. Baldwin (1983) 415.

144 See Dix/Houston (2006) and Nicholls in this volume.

145 See Nicholls (2011) 135–37, for a specialist catalogue associated with the Palatine Library. See also the discussion by Nicholls in this volume.

146 Blum (1991) 95–123, 226–43, rightly emphasises the central importance of the Alexandrian Library to Callimachus' project.

Sean A. Adams

Translating Texts: Contrasting Roman and Jewish Depictions of Literary Translations

The Greek and Roman worlds were polyglot, and although Greek and Latin were dominant in particular regions (East and West, respectively), this impression of linguist uniformity hardly hid a complex linguistic terrain. Indeed, across the Hellenistic and Roman Empires local languages continued to interact and shape the way that the dominant language users engaged with their conquered others. This political and cultural conquest resulted in the need to develop new methods of engagement and communication, both oral and written. Setting aside the act of oral translation, for which we have very little surviving evidence (although we know that it took place frequently),[1] we find that discussion of text translation was not prominent. This is not to claim that translation was not a recognised element of ancient culture, but that it was so common that it did not warrant comment.[2] Literary texts, however, were translated far less frequently and undertaken with a particular purpose(s) in mind.

In this chapter I will evaluate and contrast the ways that Roman and Jewish writers articulated their understanding of literary translation. This comparison is valuable as translators from both cultures engage with Greek literature, but do so from very different perspectives and positions. Accordingly, in order to have a fuller understanding of literary translation in antiquity, one must compare the different ways and purposes for translation that stem from both dominant and non-dominant peoples. It is this comparison that provides substantial insights into the nature of literary translation and highlights specific elements that are distinctive to particular groups. This chapter does not only look at specific literary translations and how the rendered text aligns with the original, although such texts will be consulted. Rather, the purpose of this investigation is to evaluate how literary translation is discussed and presented by authors with their intentionality and purpose being of primary importance.

We will begin by briefly looking at some school texts that give insight into the training a student received with regard to translating texts. Following this we will assess the expressed purposes and functions of translation in the Hellenistic and Roman eras with a particular eye towards how Roman and Jewish authors engaged with Greek literary culture. I will outline some specific similarities and differences

1 On the important difference between 'interpreting' and 'translating', see Feeney (2016) 32–40. For an example of a misunderstanding between Romans and Greeks over the difference between *fides* and πίστις, see Polybius, *Hist.* 20.9–10; Livy, *Hist.* 36.28. Cf. Caesar, *Bell. gall.* 1.19; Sallust, *Bell. Jug.* 109.4; Gellius, *NA* 6.14.7–9.

2 Janse (2002) 334; Rajak (2009) 138.

Sean A. Adams, University of Glasgow

https://doi.org/10.1515/9783110660982-010

and argue that different motivations and expectations for translating works, especially with regard to power relations, are expressed in how select authors are said to have approached the task of translation. In particular, we find differences with regard to directionality, religiosity, fidelity, and chronology.

Translation in Ancient Education

In order to create successfully a new translation of a text from a source language to a target language the translator required a certain level of literary training in both languages and writing systems.[3] Although it might not appear as such, learning the skill of translation was of fundamental importance within the school curriculum. In most cases, it was not rendering one language into another, but translating elements of ancient Homeric Greek into current vernacular. As widespread knowledge of Homeric (as well as Ionic and Attic) Greek receded into the increasingly distant past, there was a growing need for translational glosses to assist the challenged reader. This led not only to the rise of ancient Greek-to-Greek lexica,[4] but also to the increased occurrence of glosses and paraphrases of difficult expressions, such as those found in D-scholia of Homer in Greek manuscripts. Accordingly, intralingual translation of Homer is a recognised element of literary education.[5] For example, a first century scroll contains a school exercise in which the student paraphrased line by line the text of *Iliad* 2.617–70 (*PSI* XII 1276).[6] Fragment 233 of Aristophanes shows a scene of oral instruction on Homeric words, providing an indication of how early this practice needed to start. Similarly, a more advanced exercise is found on a set of wooden tablets in which the student provided a rhetorical retelling of *Il.* 1.1–21.[7] Such evidence reveals that intralingual *translation* was an integral part of schooling in the later Graeco-Roman world.

3 E.g., *P.Fouad* I 5 (4th-5th century CE) is a codex that presents Virgil, *Aen.* 3.444–468 with a Greek translation on the facing page. The hand, in both languages, is a practiced cursive. For the recent claim that a person competent in two languages would likely have good handwriting, see Mugridge (2016) 30.

There are a few mentions of ancient authors learning a second language later in life, but only with difficulty (E.g., Athenaeus, *Deipn.* 3.127b; Lucian, *Merc. cond.* 23; Josephus, *Ant.* 20.263–64; Plutarch, *Dem.* 2.2). On the latter, see Mossman (1999). For the need for bilingualism in certain Egyptian archives, see Clarysse (2010).

4 For a larger discussion, see Tosi (2015) 1.622–36.

5 Morgan (1998) 110–19; Cribiore (2001) 194–97. Makrinos' judgment is apt: 'Homeric paraphrases are a long and continuous link in the history of the literary and linguistic tradition of the Homeric text'. Makrinos (2011) 2.626.

6 For a good example of paraphrasing in antiquity, see Plato, *Resp.* 393d-394a on the *Iliad*. For important differences, see Theon, *Prog.* 108–110 (Patillon); Quintilian, *Inst.* 10.5.4–11.

7 Spooner (2002) 25–26.

There is little evidence, however, that interlingual translation was part of the school curriculum in the Hellenistic and early Roman eras. This is not particularly surprising as the challenge of learning one language is difficult enough without adding a second. As far as ancient teachers were concerned, a youth would have only learned to read and write one language at a time: if one was exposed to Latin at home, than it would be best for that student to learn Greek at school.[8] In *Gramm.* 1.1, Suetonius mentions that Livius and Ennius gave instruction in Rome, but did no more than translate (*interpretabantur*) Greek poets and give readings of what they had composed in Latin.[9]

Our best example of dual language use in the classroom comes from the *Hermeneumata Pseudodositheana*, anonymous bilingual instruction manuals that were used to help teach Latin to Greek speakers. Dating from the first century CE, the *Hermeneumata* contain two types of glossaries (alphabetic and thematic) of approximately 30,000 words and a series of dialogues which describe typical events in a student's day (cf. the glossaries by Ps.-Philoxenus and Ps.-Cyrillus).[10] The *Hermeneumata* are not the only bilingual school texts to survive.[11] Another example is that of *O.Max*.inv. 356, a first/second-century, bilingual abecedary in which the Greek reader is informed how the Latin letters are to be pronounced.[12] The *Hermeneumata* and other surviving compositions (e.g., *P.Oxy.* LXXXII 5302) indicate a growing focus on speaking rather than on the act of literary translation, but provide evidence that dual language acquisition was of concern to some communities. Although they represent a later development of the education system, these examples likely provide an indication of bilingual teaching methods of the Hellenistic era. The dearth of earlier, identifiable school texts likely indicates that a person learned to write their second language in a different setting, such as a different language school or at a later stage of their education (e.g., *UPZ* I 148).[13] Despite this lack of primary evidence for interlingual translation of literary texts in the school system, ancient authors did undertake the task of translation, although exactly what this entailed is not easy to determine.

The recognition that elements of translation were included in the education system bolsters the idea that translation was a regular and important task in antiquity.

---

8 Quintilian, *Inst.* 1.1.12; Augustine, *Conf.* 1.13–14. By the end of the Republican period the Latin translation of Homer by Livius Andronicus was being used in Latin education (Horace, *Ep.* 2.1.69–71; cf. Suetonius, *Gram.* 9). Cf. Cribiore (1996) 9, 148.

9 Cf. Nepos, *Gramm.* 4.2 = F61 (Teubner).

10 Gayraud (2010) 35–44; Biville (2002) 84–85; Kramer (2001) 30. For a new introduction, translation, and commentary, see Dickey (2012–2015) esp. 1.50–52.

11 Aesop's *Fables* and Hyginus' *Genealogy* are also thought to have been used in class settings. Cf. Marrou (1956) 263–64.

12 Cf. *P.Ant.* I fr. 1; *P.Oxy.* X 1315. For the appropriation of Latin teaching materials for Greek speakers, see Dickey (2010). For additional evidence of Greek-Latin learning, see the table in Dickey (2012–2015) 7–10.

13 Cribiore (1996) 30.

This claim is not surprising, but by framing our discussion of translation in educational terms we see that the foundation for literary translation was laid early in a person's training. Indeed the exposure to Greek authors and literary models is an essential element for the claim by Romans that its literature is a continuation of Greek culture.[14]

Roman and Jewish Translations

Although a definition of 'translation' can be readily provided by most people, the principles of translation are difficult to articulate and deceptively complex. The process of rendering an element from a source language into a receptor or target language is challenging and there is substantial debate over how these terms should be understood and how this processed should be engaged.[15] Cultural and temporal distances pose an even greater challenge to scholars of antiquity; we conceptualise and define translation in (sometimes vastly) different ways than the ancients did. In the following sections we will investigate the ways Roman and Jewish authors present their understanding of translation of literary texts.[16] As representative authors describe specific acts of translation we will notice points of similarities and differences in approaches as well as rationales for engaging in this practice.

Translations by Latin Authors

The Romans were not the first people in their vicinity to engage with the practice of translation, but were preceded by their Greek neighbours. An early example of this comes from Herodotus, who regularly tells of tales from distant lands, such as those of the Egyptian kings.[17] For instance, Herodotus recounts the story of an unnamed man who stole from the king in such an ingenious way that king Rhampsinitus offered him his daughter in marriage (*Hist.* 2.121). Herodotus uses the language of speaking, rather than of reading (e.g., ἔλεγον), when describing these tales, but such examples show that foreign stories were beginning to be included within Greek literature and that translation of the tale must have taken place at some level.[18]

14 E.g., Ovid, *Amor.* 1.15. Cf. Feeney (2016) 4–13.

15 On the difficulties of defining 'translation' and the risks of imposing modern ideas onto ancient authors, see Hermans (2002) 10–23. Indeed, all of the essays in that volume contribute to this discussion.

16 Brock rightly notes that the type of document influences the nature of translation, with legal texts tending to be more 'literal' and literary text translated more 'freely'. Brock (1972) 17.

17 See also the tales of Sesostris (*Hist.* 2.102–104, 107–10) and Sethon (*Hist.* 2.141).

18 Cf. Stronk (2011). For some additional references to translation in Greek literature, see Herodotus, *Hist.* 7.87; Thucydides, *Hist.* 4.50; Luke 23:38.

One of the earliest authors to discuss the relationships between different languages, both written and spoken, was Aristotle, who claims in the opening of his *Interpretation:*

> Ἔστι μὲν οὖν τὰ ἐν τῇ φωνῇ τῶν ἐν τῇ ψυχῇ παθημάτων σύμβολα, καὶ τὰ γραφόμενα τῶν ἐν τῇ φωνῇ. καὶ ὥσπερ οὐδὲ γράμματα πᾶσι τὰ αὐτά, οὐδὲ φωναὶ αἱ αὐταί· ὧν μέντοι ταῦτα σημεῖα πρώτων, ταὐτὰ πᾶσι παθήματα τῆς ψυχῆς, καὶ ὧν ταῦτα ὁμοιώματα πράγματα ἤδη ταὐτά.
>
> Spoken words are symbols of impressions of the soul; written words are the signs of words spoken. As with writing, so also with speech, it is not the same for all races of people. But the mental impressions themselves, of which these words are primarily signs, are the same for the whole of humanity, as are also the objects of which those impressions are representations or likenesses, images, copies (*Int.* 1, 16a4–8).[19]

Aristotle recognises the fact that different people groups have different languages and that this dissimilarity is expressed both in word and in script. However, according to Aristotle, the mental impressions of the reality of the world are the same for all people, but are only expressed using conventions that are appropriate to the language of the observer. In this statement, Aristotle asserts that mental impressions and their objects are secure and stable despite the fact that they are viewed by people from different ethnicities and languages. Although not explicit, Aristotle implies that at some level words from differing languages can be equated because they are merely signifiers referring to the same fundamental impression. If one could find the appropriate signifier in the target language, then, according to Aristotle's understanding, one would be able to transfer his or her idea with minimal loss or corruption of meaning because the true nature of the signified is identical in both languages. This statement by Aristotle is brief and, unfortunately for us, Greek authors rarely remarked on the nature of interlingual textual translation, nor was the practice prominent in their literary activity.[20] As will be discussed further below, this is likely due to the dominant position and high prestige that Greek language and literature held in antiquity. Nevertheless, Roman practice marks an important shift in how translations were viewed and understood in antiquity.

Roman authors were much more interested in literary translation than their Greek neighbours, discussing its nature, purpose, and function for centuries.[21] In

---

19 Adapted from Cooke/Tredennick (1938). Cf. Aristotle, *Poet.* 1457b1–5.

20 Cf. Most (2003) 389. Certain texts (e.g., *senatus consulta* documents) were translated literally from Latin to Greek, but this direction of translation was not common for literary texts in the Hellenistic and early Roman era. Cf. Brock (1979) 74. Later, Greek translations of Virgil become common, especially as school texts/exercises. E.g., *P.Ryl.Gr.* 478a-b. The *Oxyrhynchus Glossary* (*P.Oxy.* XV 1802 + LXXI 4812) does show that Eastern languages (i.e., Persian, Babylonian, and Chaldean) were transliterated and glossed for Greek readers. Cf. Schironi (2009). See also the discussion of lexica in Hatzimichali's chapter.

21 For important discussions of the development of Roman translation theory that takes into account the historical situation of the authors, see McElduff (2013); Feeney (2016).

particular, Roman authors throughout the late Republic and early Principate regularly engaged with Greek literary culture, and as early as the third century BCE we find examples of Homer's works being translated into Latin.[22] The first individual reported to have attempt this was Livius Andronicus (ca. 284–204 BCE), who in addition to appropriating Greek dramatic works for Latin consumption, rendered Homer's *Odyssey* into Saturnian verse, complete with latinising names for the divine pantheon.[23] Livius was not the only person to attempt this feat; Gnaeus Matius and Ninnius Crassus are also said to have translated Homer (Gellius, *NA* 7.6.5), and Seneca reports that a subsequent translation was attempted by Polybius, a freedman of Emperor Claudius (*Polyb.* 11.5).[24]

Roman translation methods could be very free.[25] The difficulty in engaging with two scrolls simultaneously, as well as the practice of working from epitomes or memory, could lead to 'translations' that would hardly be worthy of the term by modern standards.[26] Such practical issues, however, do not require a free approach to translation. Rather, the author's (and wider culture's) view of the original text and the purpose for engaging in translation are more important factors in this regard. To the Latin translator, the Greek source text was not considered sacrosanct, but could be heavily abridged (e.g., Livius Andronicus) and blended with new material penned by the translator cum author (e.g., Germanicus Caesar's translation of Aratus' *Phaenomena*).[27] Such an approach to translation—namely not being too beholden to the original, but trying to improve upon it—is explicitly endorsed by Pliny (*Ep.* 7.9.4) and Horace (*Ars* 131–35). These comments support the view that the Roman translators were deeply and unavoidably influenced by their personal, social, and cultural histories, which shaped the way they approached the translator's task.[28] Not only are

---

22 Although we will be focusing on literary works, we must also recognise the reality of legal and official translations that were very much part of the wider ancient world (e.g., Rosetta Stone, Augustus' *Res Gestae*, senatorial decrees, etc.). For a recent discussion of how the language of each inscription is tailored to its perceived audience, see J. Larson (2011) 50–61.

23 See Courtney (1993) 45–46; Armstrong (2008) 170, 174–75.

24 Not all translations of Homer were thought to be of a good standard and Attius Labeo's 'literal' version of both poems was heavily satirised by Persius (*Sat.* 1.4–5, 50). Conversely, Baebius Italicus's *Ilias Latina* recounted the whole of the *Iliad* in one book and became the standard Latin translation of Homer in the West through the Middle Ages. Cf. Scaffai (1994).

Homer was not the only Greek epic poet translated for Latin readers: P. Terentius Varro Atacinus (b. 82 BCE) made a translation of Apollonius' *Argonautica* (Quintilian, *Inst.* 10.87). Courtney (1993) 238–43. Cf. Terence, *Eun.* 7–8.

25 There is much literature on the definitions of 'free' and 'literal' with regard to translation. In this chapter, I am not using them technically, but to approximate the relative relation between source and rendered texts. For a more thorough discussion and problematisation of this concept, see Barr (1979); Davila (2005).

26 Cf. Cicero, *Att.* 2.6.1; Pliny, *Ep.* 3.15.17; 7.9.1–16.

27 Gellius (*NA* 1.3.11–12) speaks of Cicero omitting and condensing sections of Theophrastus' *On Friendship* in his 'translation'.

28 See the foundational discussion of Siméoni (1998) esp. 32.

translators constructs of their society, they are at the same time constructors of the inherited wisdom of future translators; they are both shaped by and are shapers of tradition. The translators in their work are constrained (to some extent) by the expectations of their readers. At the same time they seek to adjust their received expectations so as to promote and achieve their own personal goals. As a result, the act of translating literature in the Roman world is highly complex and the translator must navigate multi-currented waters if he or she is to succeed in his or her endeavours.

Translation can also be seen as a tool of the conquerors and of the powerful. Rome, in their subjugation of Greece, not only took their freedom and wealth, but also repurposed their literature to tell a new, Roman story. Romans were aware of the cultural superiority of the Greeks and many of the authors of surviving Lain texts express a deep dissatisfaction at this imbalance and a desire for Latin language and literature to be elevated so as to (at the minimum) become a peer to Greek.[29] This understanding is well expressed by Crassus in Cicero, *De or.* 1.155:

> Postea mihi placuit, eoque sum usus adulescens, ut summorum oratorum Graecas orationes explicarem, quibus lectis hoc adsequebar, ut, cum ea, quae legeram Graece, Latine redderem, non solum optimis verbis uterer et tamen usitatis, sed etiam exprimerem quaedam verba imitando, quae nova nostris essent, dum modo essent idonea.
>
> Afterwards I resolved—and this practice I followed when somewhat older—to translate Greek speeches of the most eminent orators. The result of reading these was that, in rendering into Latin what I had read in Greek, I not only found myself using the best words (and yet familiar ones), but also coining by analogy certain words such as would be new to our people, provided that they were appropriate (trans. Sutton and Rackham).

Translation for Crassus, in addition to developing his oratorical skills, also becomes a civic duty to one's nation, a way to benefit the Roman people.[30] Similarly, Cicero's critique of his countrymen who prefer Greek literature over Latin also addressed the necessity of translation for the development of Latin literary culture (*Opt. gen.* 18). More importantly, these translations will wrest control from Greek authors and place it firmly in the hands of Rome, eliminating the dependence on Greek libraries and authors (Cicero, *Tusc.* 2.5–6) and establishing a new benchmark for Latin reading (*Opt. gen.* 23). This requires that the translator add his own voice to the newly-formed work and does not translate the text overly literally; 'for it is easy to translate (*interpretari*) thoughts, that I could do if I did not wish to be myself' (Cicero, *Leg.* 2.17).[31] Fidelity to the text, therefore, does not become the primary purpose of

---

29 Cf. Cicero, *Tusc.* 1.3–6; *Acad.* 1.4–5, 10; *Leg.* 1.5–6.

30 McElduff (2013) 10. Cf. Cicero, *Fin.* 1.4–6, 10; *Acad.* 1.5. For a detailed discussion of Cicero's practice of introducing Latin glosses to Greek philosophical terms (e.g., *globus* for σφαῖρα), see Sedley (2012) 189–92.

31 At times, Cicero claims to make literal translations (e.g., *Tusc.* 3.37; *Div.* 1.61, 'I have reproduced Plato's very words [*verba ipsa*]'.), although his closeness to the original text is questionable (cf.

Latin translation as it would render the translator mute and so perpetuate the voice of the Greek original.[32] Rather, translation among Latin authors becomes a method of control and rebranding, allowing the original to say only what they wished it to say (e.g., Cicero, *Off.* 1.6). Indeed, to be too faithful to the text would undermine the intention of translation, namely to imbue previous Greek thought with Roman ideals and to provide a new voice for the Roman populous.[33]

This move away from the strict translation of the original towards a creative repurposing is highlighted by the attribution of a new authorial name to the work. These Latin translations are not to be done anonymously, as many translation works were in antiquity. Rather, they signify their compositional stamp by actively claiming the 'new' work as their own. Furthermore, reattribution is found in a range of literary genres (e.g., epic, tragedy, comedy, philosophical texts, etc.), which helps distinguish this practice from the rote translations of legal and other documentary texts.[34] Accordingly, Latin translation becomes another form of conquest, whereby formerly Greek accomplishments are co-opted and become part of Latin literary culture.[35]

One important consideration for these discussions by Latin authors is the assumption on their behalf that the readers of Latin translations would have access to the original, source-language text. As members of the Roman educated elite would likely have also read the original text, the translator was not compelled to replicate the original, nor was that his primary purpose for engaging in translation ac-

---

Fin. 3.15). For Cicero's famous comparison between word for word translations and those engaged by orators, see *Opt. gen.* 14; Jerome, *Ep.* 57.5.

32 On the other hand, it does appear that Cicero thought that some instances of translation required greater fidelity to the original (e.g., *Fin.* 1.7), but here in the tradition of Ennius and Afranius it still allowed for variation. Cicero's close adherence to a source text is found most prominently in the surviving fragments of his translation of Plato's *Timaeus*, but, as far as we are aware, Cicero still felt free to add glosses and to omit parts of the text (e.g., the entire conversation preceding Timaeus' speech and Timaeus' opening prayer). For a full list of Cicero's translations from Plato, see Powell (1995) 279–80.

However, note the claim by Terence (*Ad.* 7–11) that he included a passage of Diphilus' comedy, *Comrades in Death*, that was omitted by Plautus and that he translated it 'word for word' (*verbum de verbo*). Unfortunately, as the Greek play is lost, we have no way of determining the veracity of Terence's claim. Cf. Dares of Phrygia, *De excidio Troiae historia* 5–10

33 Not all translations do this. Cicero laments that poor translations, made from inferior Greek texts, give Latin philosophical writings a bad name (*Fin.* 1.8; cf. *Fam.* 15.19.2).

34 E.g., Attius Labeo (Courtney [1993] 350); Plautus, *Bacchides*; Cicero, *Off.* 1.6. For a named Latin translation of Aesop's *Fables* (by Quintus Ennius) complete with additional Latin moralisation, see Gellius, *NA* 2.29.20.

35 One potential counter example to this would be the need by Greek authors to use Latin sources for their composition. One instance of this would be Plutarch's use of Livy in *Marc.* 11.4; 24.4; 30.4. Thanks to Elizabeth Corsar for raising this point. For a foundational discussion of the use of Latin in the Greek East, see Rochette (1997).

tivities.[36] Rather, the translator would be judged by his ability to make creative use of his model and through this adaptation his own originality and skill would be expressed.[37] This textual duality and synoptic reading practice fundamentally changes a perceived rationale for engaging in translation, namely, to make the source text available to a readership to which it was previously excluded.[38]

One final point is needed to be highlighted prior to turning our attention to Jewish authors, namely that all of the above mentioned author are elite members of Roman society. Although this is not surprising, given the way that literature is written and preserved from antiquity, we need to be aware that social issues are likely also at play. Cicero's (and other's) comparison between orators and 'interpreters' (*interpres*) embed class distinctions (e.g., *Opt. gen.* 14; *Leg.* 1.14.9; *Rhet. Her.* 2.16.11) and reinforce, not a particular type of translation practice, but the hierarchy of who is worthy to bring Greek literary works into the Latin fold.[39] Such a perspective has at its heart the struggle for power and control, not only within Latin culture, but also how Latin literary culture engages with its Greek conquests.

This brief discussion of Roman translation practice and debates provides an important counterpoint to how Jewish authors, who are political and cultural minorities, understand and describe their act(s) of translation.

Jewish Authors and Translation

Jewish scholars were active translators throughout the Hellenistic and Early Roman eras.[40] Although a lot of translated works, such as the Septuagint and Sirach, are thought to be translated in Egypt,[41] there is evidence that translation into Greek

---

36 So Marrou (1956) 255.

37 E.g., Cicero, *Acad.* 1.8. Cf. Bassnett (2014) 55.

38 For examples of comparison of Greek and Latin texts, see Gellius, *NA* 2.23; 13.31.17 (cf. Cicero, *Nat. d.* 2.91); 17.10 (comparing Pindar, *Pyth.* 1.21–26 and Virgil, *Aen.* 3.570–577).

39 McElduff (2009). Copeland rightly highlights that this distinction also encodes an expectation of one's level of education and thus disciplinary boundaries (e.g., Cicero, *De or.* 1.187). Cf. Copeland (1991) 16–30.

40 For the different translations of the LXX over this time period, see Fernández Marcos (2000) 109–87. For later Jewish translation endeavours, see de Lange (2015).

Jews were not the only people group to translate texts into Greek in the Hellenistic Era. For example, we have evidence that the Demotic Manual of law was translated into Greek (cf. *P.Oxy.* XLVI 3285) as well as the *Oracle of the Potter* (*P.Oxy.* XXII 2332). Similarly, select pieces of Egyptian literature have also been rendered into Greek. *The Dream of Nectanebo* and the *Tale of Tefnut* both are thought to be translated works, and the Coptic version of *Tale of Tefnut* and the Demotic version of the *Dream of Nectanebo* survive. *Story of Sesonchis* might also be considered a translated work. Fraser (1972), 1.676; Cf. West (1969); Rutherford (2016) esp. 5–16.

41 E.g., Sirach, *Praef.* 27; *Letter of Aristeas, passim.* Cf. Joosten (2012) 185–239. On translations of legal texts, including the Septuagint and the Demotic Case-Book, see Mélèze-Modrzejewski (1995) 1–19.

also occurred in the heart of Judaea.[42] For example, the colophon to Greek Esther states that it was 'translated by Lysimachus son of Ptolemy, a member of the Jerusalem community' (καὶ ἑρμηνευκέναι Λυσίμαχον Πτολεμαίου τῶν ἐν Ιερουσαλημ, 10:3 l, F11). Although this declaration of Jerusalem translation is distinctive, this act of translation is not unique as recent studies have found that the linguistic situation in Roman Judaea was diverse and more complex than previously acknowledged.[43]

Translating literary works from Hebrew or Aramaic into Greek, some Jewish translators explicitly acknowledged the challenges of translation and the inability of the target language to capture the nuance and force of the source language.[44] For instance, the translator of Sirach, in the preface to his grandfather's work (*Praef.* 15–26), declares that his work, as well as those that have gone before, are not a little different from the original (οὐ μικρὰν ἔχει τὴν διαφορὰν ἐν ἑαυτοῖς λεγόμενα, *Praef.* 26).[45] In this case the author was not lamenting the inability of the target language to capture the meaning of the source language. Rather, the apology is that by adhering closely to the structure and meaning of the Hebrew the force and impact of the original is lost in Greek: 'For those things originally in Hebrew do not have the same force when rendered in another language' (οὐ γὰρ ἰσοδυναμεῖ αὐτὰ ἐν ἑαυτοῖς Εβραϊστὶ λεγόμενα καὶ ὅταν μεταχθῇ εἰς ἑτέραν γλῶσσαν, *Praef.* 20–21).[46]

Of primary importance for this paper is the *Letter of Aristeas* and its literary depiction of the translation of the Septuagint.[47] It is rare for a work from antiquity to have translation as a major theme, which has led *Aristeas* to be the focus of a number of scholarly studies.[48] Set in the Alexandrian royal courts, the opening and closing of

---

42 Recently, Aitken has provided a detailed study of Jewish and Egyptian documentary translation techniques, highlighting similarities between surviving Demotic documents and the Septuagint. See Aitken (2016). For a nuanced consideration of the LXX in light of Greek inscriptions in Egypt, see Aitken (2014).

43 E. g., the Greek scroll of the Minor Prophets at Nahal Hever (8HevXIIgr) and the Greek fragments found at Qumran: 4Q119; 4Q120; 4Q121; 4Q122; 7Q1; 7Q2. Cf. Hengel (1974). This is most recently argued by Wise (2015).

44 Such a literary apology is rarely a concern for Latin literary translations. So Fewster (2002) 232; Rajak (2009) 136. However, cf. Gellius, *NA* 17.20.8.

45 Ziegler (2016).

46 Cf. Iamblichus, *Myst.* 7.5; Seneca, *Polyb.* 11.5; Gellius, *NA* 11.16. See also Aitken (2011). For the view that the apology is for the poor quality of the writing, as opposed to deviation from the Hebrew, see Wright (2003) 11–20.

47 Shutt's translation of *Aristeas* 9–10 implies that other texts, in addition to the Jewish Scriptures, were 'purchased and translated' (ποιούμενος ἀγορασμοὺς καὶ μεταγραφάς) for the Alexandrian library. Although this translation is possible, it is better rendered 'purchased and transcribed' as §11 and 15 make it explicit that the Hebrew text needs translation (ἑρμηνείας προσδεῖται). Cf. Shutt (1985) 12.

48 Honigman (2003); Niehoff (2011) 19–37; Matusova (2015); Wright (2015). For the edition used, see Pelletier (1962) 100–240.

Aristeas tells the story of the 'translation' of the Hebrew Scriptures into Greek (§§9–11, 301–321) and recalls all of the pomp and ceremony surrounding this process. The text is written to Philocrates (§1) from the perspective of Aristeas, a Greek administrator reported to be involved in the events. Both characters are Greek in ethnicity and so provide an 'outsider's' perspective to the events. The text, however, is Jewish in origin, written by an unknown author likely in the second century BCE.[49] The proposed narrative rationale in *Aristeas* is that Philocrates has a deep love of learning (κατειληφὼς ἣν ἔχεις φιλομαθῆ διάθεσιν, §1), and so would be interested in Aristeas' experiences and this Jewish text.[50]

In light of the work's literary setting in Alexandria, there has been a move recently to read *Aristeas* as part of Alexandrian literary culture and to compare the contents of the text to then contemporary Greek methods of textual scholarship.[51] One of the first indications of this perspective is the use of specific scholarly vocabulary in §§30–32:

> Τοῦ νόμου τῶν Ἰουδαίων βιβλία σὺν ἑτέροις ὀλίγοις τισὶν ἀπολείπει· τυγχάνει γὰρ Ἑβραϊκοῖς γράμμασι καὶ φωνῇ λεγόμενα, ἀμελέστερον δέ, καὶ οὐχ ὡς ὑπάρχει, σεσήμανται, καθὼς ὑπὸ τῶν εἰδότων προσαναφέρεται· προνοίας γὰρ βασιλικῆς οὐ τέτευχε. [31] Δέον δέ ἐστι καὶ ταῦθ' ὑπάρχειν παρά σοι διηκριβωμένα, διὰ τὸ καὶ φιλοσοφωτέραν εἶναι καὶ ἀκέραιον τὴν νομοθεσίαν ταύτην, ὡς ἂν οὖσαν θείαν...[32] Ἐὰν οὖν φαίνηται, βασιλεῦ, γραφήσεται πρὸς τὸν ἀρχιερέα τὸν ἐν Ἱεροσολύμοις, ἀποστεῖλαι τοὺς μάλιστα καλῶς βεβιωκότας καὶ πρεσβυτέρους ὄντας ἄνδρας, ἐμπείρους τῶν κατὰ τὸν νόμον τὸν ἑαυτῶν, ἀφ' ἑκάστης φυλῆς ἕξ, ὅπως τὸ σύμφωνον ἐκ τῶν πλειόνων ἐξετάσαντες καὶ λαβόντες τὸ κατὰ τὴν ἑρμηνείαν ἀκριβές, ἀξίως καὶ τῶν πραγμάτων καὶ τῆς σῆς προαιρέσεως, θῶμεν εὐσήμως. Εὐτύχει διὰ παντός.
>
> The books of the law of the Jews (with some few others) are absent from the library. They are written in the Hebrew characters and language and have been carelessly transcribed and not as they should be, according to the reports of experts, for they have never received royal patronage. These [books] should also be for you in an amended form since this legislation is both most philosophical and pure, inasmuch as it is divine... If it pleases you, O king, a letter shall be written to the High Priest in Jerusalem, to send men who have lived noble lives and are mature, skilled in their law, six from each tribe, so that after the examination of the text and the agreement of the majority, and receiving an accurate interpretation/translation we may place it conspicuously [in the library], worthy of the deed and your purpose.

In this passage Aristeas emphasises the need for well maintained and curated texts that come from an accurate source text. Ancient scholars were aware that not all texts were equal in quality and that the best texts were regularly held in the originating city's archive/library.[52] In this case, the focus on Jerusalem aligns well with the

---

49 Wright (2015) 21–30.

50 The reader is also thought to share Philocrates' love of learning (§§7, 171, 300) and interest in the Jewish people, religious topics, as well as the ultimate importance of piety (§§2, 5).

51 So Honigman (2003); Niehoff (2011) 19–37; Matusova (2015).

52 E.g., the texts of Sophocles, Euripides, and Aeschylus held at Athens. Cf. Galen, *Commentary on the Epidemics of Hippocrates* 2.4, Kühn (1821–1833) 17.1.607–608; Gellius, *NA* 1.7.1. For a discussion of the importance of the author's personal copy, see Gellius, *NA* 1.21.2.

scholarly practice at Alexandria to secure the best text editions for study. The previous text, its exact nature is left unspecified, is said to be ἀμελέστερον and καὶ οὐχ ὡς ὑπάρχει, σεσήμανται, both serious and detrimental accusations against the nature of the current text.

The phrase ἑρμηνείαν ἀκριβές also has scholarly overtones, but with important ambiguity as it expresses exactness, either to '(scholarly) interpretation' or to 'translation'.[53] The recurring use of the term ἑρμηνεία in *Aristeas* (§§3, 11, 32, 120, 301, 308) is one of the distinguishing features of this narrative as the author describes the work of the Jewish translators using this term and not the more explicit μεθερμηνεύω, 'to translate'. The term μεθερμηνεύω is used in §38, but it is in the letter from the king (§§35–40) and so does not represent the viewpoint of Aristeas or the Jewish scholars within the text. A similar ambiguity is found in the use of μεταγραγή, 'transcription' (§§9, 10, 15, 45, 46, 307), which could also imply translation, but does not have sufficient clarity within the narrative.[54]

In the final section of *Aristeas* (§§301–307) we return to the theme of translation and in §302 the author provides a description of the process undertaken by the Jerusalem scholars: 'and they accomplished it, making each detail agree by comparison with each other (σύμφωνα ποιοῦντες πρὸς ἑαυτοὺς ταῖς ἀντιβολαῖς)'. Ἀντιβάλλειν is a technical term for the collation of manuscripts and is coherent with the expressed desire to have an accurate text earlier in the narrative. This is followed by the reading aloud (ἀνάγνωσιν) and the interpretation (διασάφησιν) of every passage (§305). This process is parallel to that described by Dionysius Thrax (*GG* I.I 5), who argued that the first step in understanding the text was correct reading (ἀνάγνωσις), which includes expression, intonation, and divisions, and was to be followed by the interpretation (ἐξήγησις) of the text.[55] For Dionysius the process concludes here, but for the Jewish authors in *Aristeas* an additional step is required, namely, rendering the now-understood text into a new target language. Matusova argues that applying Greek grammatical theory to the translation of a work is an 'unprecedented adaptation' and has strong implications for ancient translation theory.[56] This is a bold claim, but not completely unwarranted, as the most striking feature of *Aristeas* is that the author does not describe the scholarly activity through translation language, but with editing vocabulary. The author of *Aristeas* could have used such terms when discussing the creation of the Greek version, but instead he employed alterative language which has strong resonances with Alexandrian scholarship in general and Aristarchus in particular.[57]

---

53 Zuntz (1959) 121–22.

54 Philo claims that the seventy should not be called 'translators' (ἑρμηνέας), but 'priests and prophets' (ἱεροφάντας καὶ προφήτας, *Mos.* 2.40)

55 van der Kooij (1998) 222.

56 Matusova (2015) 84. See most recently, Matusova (2018).

57 Montanari (2015) esp. 1.86–89; Wright (2008).

What is important to recall here is that the account in *Aristeas* is not to be taken as a straight, unadulterated description of events that actually happened, especially in light of the strong difference between the LXX and the *Aristeas* narrative.[58] This is not to say that the narrative is devoid of historical facts, or that it could not be used for historical reconstruction, but that the text is inherently ideological and that the author had a specific communication goal in mind.[59] The discussion of the 'translation' in terms of text-editing vocabulary is a strategic move on behalf of the author, one that employs current ideas and methodology and transfers them back to the time of Ptolemy II.[60] The publication of Aristarchus' edition of Homer around the time of *Aristeas*' composition represents the pinnacle of Alexandrian text criticism and set the standard for future editions of texts. It is questionable whether or not the original translators employed grammatical methodologies; however, it is apparently important to the author of *Aristeas* that the Jewish translators appeared to have done so. The use of Greek scholarly vocabulary, therefore, frames the discussion and depiction of Greek Scripture and is employed by the author to show that the translators adhered to the best of Greek scholarly practice. Moreover, this practice was almost exclusively used for literary works, and so in making this parallel the author imputes a literary quality to the work.[61]

One of the more obvious suggestions for the adoption of this presentation was the desire to present the Jewish scholars, especially those outside of Alexandria, as knowing and using advanced reading techniques. In the contest of cultures the author of *Aristeas* is making a statement regarding the Jewish people and their level of literary sophistication; the seventy-two Jewish scholars from Jerusalem are not under-equipped for the task, nor is their city a backwater metropolis lacking proper education.[62] A major theme of *Aristeas* is the fact that the Hebrew text and its translators come from Jerusalem;[63] these translators are fully conversant with the current trends of scholarship and are able to excel in such tasks. Indeed, the fact that these scholars are named (§§47–51; cf. Josephus, *Ant.* 12.57), a practice

---

58 Cf. Wright (2006) esp. 59–61.

59 For a less complementary position, see Zuntz (1959) 123, who states, 'He who strives to elicit historical facts from this narrative is building on foundations very much less solid than sand'. Cf. Janowitz (1991).

60 Contra Zuntz (1959) 122, who writes 'however abstruse it may seem that he should have chosen to represent the translators' task as a work of critical scholarship'.

61 In addition, Ps.-Aristeas presents the original Hebrew text as a philosophical work and not only a legal text (§31) and the translation being undertaken by Jewish philosophers and not scribes/interpreters (§235).

62 On the parallel between the 72 Jewish translators and the 72 editors of Homer commissioned by Pisistratus, see Matusova (2015) 54–61.

63 Although unstated, the need to import translators and the high praise given to them implies a deep inadequacy of the Jews currently living in Alexandria and in Egypt more broadly. This motif is in tension with the claims by a number of scholars that the historical translators of the LXX hailed from Alexandria. E.g., Swete (2000) 20–21; Dines (2004) 33; Honigman (2003) 97.

that is notably absent in other surviving Jewish-Greek translations (including the Septuagint), suggests that the author of *Aristeas* was attempting to raise the literary profile of the Greek text, thus implying literary aspirations. Although this is certainly what the text presents, it is not the full explanation of the author's motivations.

The author of *Aristeas* also makes declarations regarding the nature of texts, namely that the earlier edition(s) was flawed and that the new text from Jerusalem is superior, both surpassing and replacing previous versions held at Alexandria. It is this authorised text that forms the basis of the translation and which lends its authority to the new version. The use of grammatical language might also suggest that, in the mind of the author of *Aristeas* and/or his contemporaries, the founding of the Greek version of Scripture was of the same magnitude as the establishment of Homer's text.[64] As Aristarchus' texts was thought to be the pinnacle of editorial achievement,[65] so too the author of *Aristeas* presents the LXX as perfectly accurate (πᾶν ἠκριβωμένως, §310).[66] This newly rendered version receives full approval by the priests, elders, and the πολιτεύματος (§310); echoing Moses' giving of the Law to the people at Sinai.[67] More importantly for our discussion is the claim that the Greek translation perfectly represents the original Hebrew. This fits with the larger prohibition expressed in §§310–11 that no further revision should be undertaken to the text. Such prohibitions against modifications are part of the process of text establishment and have strong parallels to the revealing of the Law at Sinai.[68] This understanding also works with the larger narrative construct and the establishment of an official version of the text in the royal library.[69]

---

64 Honigman (2003) 120; Honigman (2007) 128. This perspective is critiqued by Matusova (2015) 79–82, who highlights the differences between Homer and LXX regarding style (poetry/prose), quality (good/poor), and prestige (high/low). Although Matusova is correct in her assessment, I think she downplays the perspective of the Jewish scholars and their view of Scripture as a possible parallel to the importance of Homer's works for Greeks.

65 Pfeiffer (1968) 210–33.

66 Similarly, Philo (*Mos.* 2.40) also highlights this perfection through the translators communing with the spirit of Moses and guided by a divine hand. Cf. Kamesar (2009) 66–72.

67 Wright (2015) 443–44.

68 Orlinsky (1975); cf. Brock (1992). Philo, *Mos.* 2.25–44 highlights the divine character of the LXX and the inspired nature of the translators (esp. 2.37).

69 As an aside, consideration should be given to Rewritten Scripture as a potential translation of the biblical text (e.g., *Jubilees*, Josephus' *Antiquities*, Ps.-Philo's *LAB*; *Genesis Apocryphon* [1Q20], 11QTemple Scroll [11Q19], and 4QReworked Pentateuch [4Q158]). Rewritten Scripture is not often categorised in this way, but it might be profitable to think about it in terms of intralingual translation. This differs from the examples of interlingual translation above, but one could argue that educated Jewish authors, in some cases, were a part of both Hebrew and Greek literary cultures and so the 'translation' from Greek to Greek would differ little from Hebrew to Greek translations in terms of accessibility. More importantly for our discussion, this type of literary practice parallels Greek to Latin conversion, as that too could be considered a form of intralingual translation, since Latin authors were as much part of Greek literary culture as they were of Latin culture. Additionally, as mentioned above, there were attempts by Latin authors to rewrite Livius' translation of Homer, an activity that

Similarities and Differences between Roman and Jewish Translations

We now turn our attention to contrasting the approaches of Roman and Jewish authors based on the above discussion. The first and most obvious similarity in our study is that both groups depicted themselves, and not others, as engaging in acts of literary translation. This is not to say that Jewish and Roman authors would not have recognised the activities of other groups, but that both are focused primarily on their own actions.

Second, the language that both Roman and Jewish authors translated to and from was Greek. Although this might appear obvious, both Jewish and Roman authors had other linguistic options. For example, Jewish authors in Egypt could have translated their texts into Demotic, but, as far as we know, this was not the case.[70] Similarly, Latin was an increasingly prevalent option, but this choice was almost universally declined. The selection of which texts to translate and the direction of translation (i.e., to or from Greek), however, is an important and fundamental difference. Latin authors, it appears, were not interested in translating their texts into Greek or making their texts readily available to other cultures in native languages.[71] Instead, all of the reported translations from the Republic and Early Empire are intended to make texts from other languages known to Latin readers, bring the literary wealth of other cultures under Latin influence through 'translation'.[72]

In contrast, Jewish (and other minority cultures) translators selected their own texts for translation into the then dominant language. As a result, in addition to the common choice of translating literary (i.e., non-technical) texts, something that was not a given in antiquity,[73] we witness a clear flow of translation activities, from those with little influence or standing towards the dominant cultural and polit-

---

parallels the Jewish practice of reworking translations for a new audience/purpose (e.g., Ezekiel Tragicus, Philo Epicus, Theodectus in *Aristeas* 316). Thanks to James Aitken for helping develop this idea. For recent discussion of the nature of Rewritten Scripture, see Bernstein (2005); Brooke (2010); Machiela (2010); Zahn (2012). For later changes to Livius work and the re-writing of it in hexameter, see Courtney (1993) 45–46.

70 Other linguistic choices were made by Jewish authors in different geographic situations. For example, the translations of the Hebrew Scriptures into Aramaic (i.e., the targumim) did not take place in Egypt, but in Judaea and further east (esp. Babylon).

71 Nevertheless, in late antiquity, there were a number of Greek translations of Virgil. Cf. *P.Gongr.* XV 3; *P.Oxy.* L 3553; *P.Vindob.*inv. L 24; L 62. A similar direction of translation is found in Syriac, in which few Syriac texts were translated into Greek, but many from Greek into Syriac. Cf. Brock (1977).

72 This is not to claim that Greek was not considered a prestige language in Rome or among the Roman literary elite; there was a complex relationship between the two languages. The Latin authors above recognise the importance of Greek, but concurrently desire to undermine it through the advancement of Latin language and literature. For the discussion of prestige with regard to language, see Hudson (1980) 32–34; Dixon (1997) 9–10. On Graeco-Egyptian translations, in both directions, and the relationship between Greek and Demotic, see Rutherford (2016).

73 On this important point, see Feeney (2016) 152–78.

ical power.[74] The directionality and selection of target language is a major difference and one that is highlighted especially by Latin authors, many of whom argue explicitly that translating Greek works into Latin not only provides a benefit to Latin readers, but also enriches the Latin language; an action that profits the entire nation. This explicit motivation is not highlighted in some depictions of Jewish translations,[75] but is expressed throughout *Aristeas*, which shows awareness by the author that this was an important consideration for cultural acquisition.[76] As a result, the translation of Jewish Scripture into Greek is presented as a cultural achievement, one that would benefit both Jews and Greeks alike.

This difference in target language further emphasises the different relationships that each group has with Greek culture and their position with regard to cultural power.[77] Although both Roman and Jewish authors recognise the importance, even perhaps superiority, of Greek culture, they take different approaches to translation based on their perceived relationship with it and the relative status of each language within their respective communities. For the Roman authors, Greek culture might be superior, but they are the dominant political power. Translation for Latin authors is a continuation of this conquest, expressing a desire for dominance. Jewish authors, on the other hand, are politically dominated, both in Egypt and in Judaea, and some scholars have interpreted the translation of Jewish Scriptures into Greek as a means of cultural survival.[78] For Jewish authors, the approach to Greek texts taken by the Romans was untenable; they could not exert power in such an overt manner. Jewish authors did engage in cultural negotiation and comparison, but more often it was achieved through the construction of alternative narratives, rather than discussions and depictions of translation.[79]

---

74 A similar practice is seen in the translation of Demotic texts into Greek. Quack (2005) 172.

75 E.g., Philo, *Mos.* 2.40; Josephus, *Ant.* 12.101–109.

76 As Greek was the language of many Diaspora Jews, it is possible that the translation of Jewish texts into Greek allowed their community to have a richer vocabulary. More importantly, Greek was recognised as an important language in Egypt with high prestige, as is indicated by the number of Jewish and Egyptian authors writing in Greek (e.g., Manetho, Aristobulus, Philo, etc.).

77 Cf. De Crom (2011) 77–87.

78 E.g., Rajak (2009). For a recent study that argues that the author of *Aristeas* is concerned with ethnic identity, see Moore (2015) 205–54.

79 Many Jewish texts written in Greek make specific claims about Jewish cultural superiority and how Egyptian and Greek cultures are actually derivative of it (e.g., Philo, *Leg.* 1.108; *Her.* 214; *Spec.* 4.61; *Prob.* 57; *Mos.* 2.12, 18–20; Artapanus, F1.1; F2.3; F3.4; Ps.-Orpheus vv. 27–28, recensions B and D; Aristobulus F3.1; F4.4; Eupolemus F1). These claims did not go without response, at least by Egyptian writers. For example, Manetho provides an alternative narrative for this debate from an Egyptian perspective, denigrating Jewish history in an attempt to undermine Jewish claims and bolster their perceived cultural authority. Specifically, Manetho (as reported by Josephus), claims that Moses (F3.37//*CA* 1.279) and the Jews (F3.20//*CA* 1.229, 233) were lepers, and that Moses was an Egyptian (F3.3//*CA* 1.250), who opposed Egyptian cults (F3.4, 12//*CA* 1.239, 244, 249), and fled to Ethiopia (F3.7–10//*CA* 1.246–47). For the connections between Artapanus and Manetho, see Freudenthal (1875) 161–62; Fraser (1972) 1.705–706; Barclay (1996), 129–30.

Another important difference is the determination of which texts were worthy of translation. The texts translated by Latin authors were all of high cultural value and primarily Greek in origin. The notable exception was when, in 146 BCE, the Roman Senate commissioned a Latin translation of Carthaginian Mago's *On Farming* despite the fact that, as Pliny notes, Cato the Elder's *On Farming* was already published (Pliny, *Nat.* 18.22–23).[80] This was a very large undertaking due to the size of the work (twenty-eight volumes) and so must have been deemed important for the Roman people.[81] This was a rare instance of official recognition given to a foreign, non-Greek work especially as the rest of the Carthaginian libraries were not thought to be worth keeping and so given to the Numidian kings (*regulis Africae*).[82] Moreover, this task was given to a group of translators (*Nat.* 18.23), a practice markedly different from the individual efforts by Latin authors discussed above.

This picture of state-funded scholarship parallels the depiction of the translation of Jewish Scriptures in *Aristeas*. In both cases the foreign text, which is taken from the respective capital cities, is presented as being valuable to the dominant nation and so is worthy of consideration.[83] The perceived significance of the text results in state-funded patronage, which includes the establishment of a team of translators (with a clear leader) and the preservation of the translated text in official archives.[84]

In light of such claims of cultural heritage, the use of Greek language for translations may not solely or necessarily be a capitulation to dominance, but may also be a practical necessity. Scholars debate the levels of Hebrew and Aramaic that survived in both Alexandria and the Egyptian *chora*. The translator of Sirach implies in his prologue that one of the reasons that he translated the Hebrew work was because he desired that the wisdom of his grandfather be available to those living abroad in Egypt (ἐν τῇ παροικίᾳ, *Praef.* 34). If the linguistic capabilities of Egyptian Jews were limited to Greek, then the translation of Jewish Scriptures into Greek could also embody a practical consideration. Cf. Tcherikover/Fuks (1957) 30–32. On the bilingual situation of Egypt, see Rochette (1996); Papaconstantinou (2010).

80 See also the commission of Pompeius Lenaeus by Pompey to translate the Greek medical treatises captured from Mithridates' library (Pliny, *Nat.* 25.5–7).

81 It was used by a number of later authors. E. g., Varro, *Rust.* 1.1.10; Columella, *Rust.* 1.1.13.

82 The only other known example of a Latin translation of a Punic text comes from Sallust, who had the geographic work of Hiempsal translated so he could use it as a source for his history (*Bell. Jug.* 17.7).

83 So Zuntz (1959) 120; cf. *Aristeas* 175–79.

84 Although D. Julius Silanus is said to be superior for this task, Pliny suggests that the translation process was undertaken by a number of individuals (*Nat.* 18.23). On the high priest, see Diodorus Siculus, *Bibl.* 40.3.6. The vital difference between the two narratives is that we have strong evidence in support of the historicity of Mago's translation, whereas, the historical veracity of *Aristeas*' narrative is suspect. One possible rationale for these parallels is that the author of *Aristeas* knew of the Mago incident and adapted it. The similarities in the date of the sacking of Carthage and the composition of *Aristeas* make this possible. A more likely explanation, however, is that the processes of state-funded scholarship were well known and that its schema was adapted by the author of *Aristeas*.

Another example of royal patronage for translation is found in the preface to Dictys Cretensis' *Journal of the Trojan War*, a second century CE work that recounts how the Phoenician text was discovered during the reign of Nero and that Nero, recognising the script, called for it to be translated

Another important difference between Roman and Jewish depictions of literary translation was the perceived importance of the original text and its accuracy. Among Latin authors there was essentially no discussion as to the quality of the texts that they were translating from Greek. This silence likely does not stem from ignorance or ambivalence to this topic, but that it (presumably) was not required since the Latin readers would have had access to a version of the original and that the translation differed (sometimes dramatically) from the original that textual accuracy was not an issue. In contrast, the author of *Aristeas* was very explicit regarding the importance of textual accuracy and that this was a needed step prior to translation.[85] This is expressed most prominently through the strong use of scholarly language throughout the text.

Related to the idea of textual accuracy is the rendering of the text into the target language. The 'free' rendering embraced by Latin writers stands in sharp contrast to the depiction of close translations of Jewish authors.[86] In connection with our discussion of education practices, Pietersma has suggested that the Hebraic nature of the Greek is a result of its origins in the classroom.[87] In his perspective, the 'interlinear' pairing of the Greek with the Hebrew created a 'crib' to understand the original. This certainly could be the case and Pietersma's parallel with Homer is interesting. He is no doubt correct that 'translations' of Homer were part of the standard classroom practice (see above). What does not necessarily follow from Pietersma's Homeric analogy is the rigid adherence to the source. Homeric exercises could result in close paraphrases, but also in paraphrases and rhetorical retellings that strayed sub-

into Greek and placed in the library. This work was later translated into Latin by a certain Septimius, as is indicated by the opening letter. For a recent discussion of this text, see ní Mheallaigh (2008) 406–14. Thanks to Matthew Nicholls for bringing this text to my attention. For further discussion of this schema, see Adams (2018).

85 The translator of Sirach did not mention this issue; however, as the grandson of the author, it may be widely thought that he had access to the/an original copy. A similar perspective could be held for the translation of Esther in Jerusalem.

86 The literalness that the Septuagint translators adopted, both in reality and in literary depictions—to the point of occasional incomprehensibility of the Greek—reflects the original document and the function of the translated work, both of which are influenced by traditional translation practices. Although some have claimed that it is the sacred nature of the original that inspired such close linguistic alignment, this need not be the case. The translation style adopted by the translator of Sirach, as mentioned in his preface (*Praef.* 24–25), is similar to that of the Septuagint. It is possible that the translator was not aware that he could translate in a different way (So Wright [2003] 20); however, it is equally possible that he adopted this approach because of his respect for his grandfather's text and the growing tradition of translating works (scripture and scripture-like included) in such manner (cf. Aitken [2016]). That the methodology applied in the creation of the LXX became influential in later translations is clear, but more work is needed to determine its exact influence.

87 Pietersma (2002) esp. 347–49.

stantially from the original.[88] Of greater importance for this study is the idea that the author of *Aristeas* describes the translation, not only in grammatical ideals, but also in terms of fidelity.[89] In doing so, the author presents the Septuagint in terms that educated readers would understand and approve, but also in a way that differs from Roman discussions.

The religious nature of the text and its veneration by the translators also influences the nature of translation.[90] Roman translators did not select sacred texts for translation; whereas Jewish Scripture was the focus of many translational endeavours. The notable exception of Latin translation of religious texts is the Vulgate, a fourth-century CE translation of Greek and Hebrew Scriptures by Jerome. This Latin translation is much later than the examples discussed above and shows clear influence of Jewish translation practices through its relationship to the Christian church. Nevertheless, the Vulgate demonstrates that not all Latin approaches to translation were the same and that preferences changed over time, especially in consideration of the nature of the source text.

Importantly, the target text, once it has been rendered, begins to act independently from its source text. This new translation, embodied by the attribution of a new authorial name, is presented as its own literary work and so garners respect or derision often apart from the way that the original text continues to be viewed. In some cases, the new text might be found inferior to the perceived excellence of the source or the translation could be viewed as superior, surpassing the text from which it was birthed (so Pliny, *Ep.* 7.9.3–6; Gellius, *NA* 11.4.1–4). In such situations, judgment comes from literary elites who have access and the ability to read both works. Indeed, the pleasure afforded to one's literary equals through a good translation was thought to be a respected leisurely pursuit and the primary audience of such endeavours (cf. Cicero, *Acad.* 1.4–5).[91] For the Jewish readers, the construction of a Greek text has important implications for the perception of the Hebrew. Honigman, for instance, claims that the blurring between editing and translating helps to prioritise the Greek text, as one only keeps the new, collated version and not the previous texts from which it was created. As a result, the Hebrew original can be forgotten now that the Greek text has been achieved.[92] Although I appreciate her perspective, I would disagree that the author of *Aristeas* advocates that view (cf. §177). Rather, the

88 What does it mean, then, that the Jewish authors felt compelled to follow the Hebrew text so closely so as to imbue the Greek with elements of the source language? Such a practice of developing 'poor' Greek was, in fact, the opposite intention of the classroom and would likely have been punished by the *grammatikos*. As a result, claiming a school setting for the origin of the LXX is still insufficient to determine exactly why the Greek translation developed as it did.

89 See also Philo, *Mos.* 2.40; Josephus, *Ant.* 12.101–109.

90 On the sacred nature, see Kelly (1979) 69. For the esteem of translated texts, see McElduff (2013) 6; Lianeri/Zajko (2008).

91 In contrast, the ridicule for a poor translation was also part of the risk of translation work. Cf. Gellius, *NA* 2.23.1–22.

92 Honigman (2003) 49.

author of *Aristeas* depicts the Greek version as being functionally identical with the original Hebrew and so can be used in proxy for the Greek speakers in Alexandria.[93] The creating of functional equivalence is markedly different from the practice of Latin authors, for whom the translated text with its new authorial stamp is intended to both supersede and displace (to some extent) the original Greek composition among Latin readers.[94]

Conclusion

In conclusion we find significant similarities and differences between Roman and Jewish depictions of literary translation. In particular, the importance of the source text, the direction of translation, and the intended purpose of the new work are key determiners of the nature of translation undertaken by the translator. A majority of the differences are a result of differences in power dynamics with Greek culture. The dominance of the Romans affords them the ability to take liberties with the text and to bring the wealth of Greek literary heritage into the Latin world. In contrast, Jewish translators did not translate Greek texts into Hebrew or Aramaic, presumably because in Egypt the Jews already had access to these texts. Rather, Jewish translators rendered their own ethnicity's works into Greek with high fidelity to the nature and structure of the original.

One of the key findings of this work is the recognition of the diversity of translation perspectives in antiquity, which I argue fall primarily along the lines of power. Engagement with another people group's literature and language engenders hierarchy and cultural negotiation; it is not a benign act. By contrasting Roman and Jewish practices, one can identify distinctive elements and demonstrate how individuals from diverse but interacting cultures can differ in the ways in which they engage in the scholarly practice of translation. Future studies, I would argue, need to engage

93 Cf. Wright (2015) 439. See also Philo, *Mos.* 2.40, which implies that comparison of the translated text with the original would yield the result of both being identical in meaning. Indeed, Philo resists seeing the Greek version as a translation, but describes it with the language of priests and prophets (οὐχ ἑρμηνέας ἐκείνους ἀλλ' ἱεροφάντας καὶ προφήτας προσαγορεύοντες), cf. *Leg.* 4.49. On the emphasis on the preservation of the text, see Veltri (2006) 36; Borchard (2012) esp. 12–15.

94 For the rhetorical claim of translational accuracy with regard to Scripture, but with clear authorial changes, see Josephus, *Ant.* 1.5, 10–13 (for explicit mention of translation model); *C.Ap.* 1.54. Cf. Josephus' description of Manetho's history in similar terms (*C.Ap.* 1.228). The literal translation, moreover, is a constant reminder to the reader that this text was not originally composed in Greek, but is a translation. McElduff (2013) 9 claims that 'Comparing a translation with its source would be difficult, even if one were interested in doing so, and we never hear of a translator doing so'. Although this might be true for Latin authors (though, see Gellius, *NA* 7.2.1–3; 8.8 [the content is sadly lost]; 10.22.3), McElduff's statement is challenged by the narrative of *Aristeas* and the history of Septuagint revisions, for which we have ample evidence that later readers were regularly comparing the rendered text with its original and engaging in new translations that would be closer to the source.

this idea when considering both translation acts and depictions of translations. In the case of *Aristeas* the presentation of translation in terms of scholarly language is a strategic move on behalf of the author in which he makes a statement regarding both the translated text and the people who made it.[95]

---

95 I would like to thank James K. Aitken and those who attended the 'Ancient Scholarship' conference at the University of Glasgow for their feedback and input.

Catherine Hezser

Rabbis as Intellectuals in the Context of Graeco-Roman and Byzantine Christian Scholasticism

In Roman and early Byzantine times various types of 'intellectuals' constructed an image for themselves in distinction to others by socialising in networks of like-minded scholars, by literary expression, and by creating 'school' traditions that continued for several generations. Philosophers, sophists, church fathers, monks, and rabbis were involved in continuous attempts to fashion their own identities and to distinguish themselves from others that claimed to possess superior wisdom.[1] Palestinian rabbinic literature of the first five centuries CE participates in this process by presenting rabbis as a particularly Jewish type of intellectual, as scholars among other scholars in the Graeco-Roman cultural environment of the Near East they lived in.[2] Palestinian rabbis' self-understanding as scholars or 'intellectuals' is not only evident in the use of the term *hakhamim*, 'sages', that equals the Greek *sophos*, but also in the envisioned settings in which scholarly encounters took place,[3] in attitudes towards earlier traditions, in the modes of compilation,[4] in the ways in which scholarly disputes are presented (see below), in rabbinic succession, and in the memorialisation of scholars through later generations of students.[5]

Before we examine similarities between Palestinian rabbis and Graeco-Roman intellectuals in more detail, it should be noted that Josephus already presents Pharisees as a 'philosophical school' (*hairesis*) alongside other such 'schools' within the Graeco-Roman context (cf. *Vita* 10–12).[6] Since he allegedly joined that 'school' him-

1 On this process of self-fashioning and boundary maintenance, see Eshleman (2012) 1: 'The right to either label could not be established once for all but had to be continually defended through assiduous self-presentation that in turn advanced implicit definitions of one's own field(s) and its rivals'.
2 On the issue of scholarship and scholasticism in antiquity, see the contributions in Cabezon (1998a). For a guide to Greek scholarly literature, see Dickey (2007). See also the contributions in Montanari *et al.* (2014), which cover the time from the classical period to the Byzantine Empire; Montanari/Pagani (2011).
3 See Hezser (2017a) 87–120 on seated sessions, and pp. 26–33 on *perambulatio.*
4 For a comparison of the scholarly processes that led to the Talmud Yerushalmi and the Apophthegmata Patrum, see Hezser (2018).
5 Reed (2008) has compared references to rabbinic succession with succession models in the Pseudo-Clementine Homilies and Eusebius. She concludes: 'Whereas Eusebius seems to pattern his understanding of succession on the lineages of Hellenistic philosophical schools, the *Homilies*' model of succession may be indebted instead to Rabbinic models' (p. 194). See also Tropper (2003).
6 For this translation of *hairesis*, see Mason (2003) 15–21. He explains that the term 'had become by Josephus' time ... a technical term for a philosophical school or sect' (p. 15 n.69).

Catherine Hezser, SOAS, University of London

https://doi.org/10.1515/9783110660982-011

self (*Vita* 10–12), this definition may represent the self-understanding of those Pharisees who, like Josephus, originated from upper-class Hellenised urban environments. As a scholarly community united on ideological grounds, Josephus compares the Pharisaic 'school' to 'the one called Stoic among the Greeks' (*Vita* 12). In his *Jewish Antiquities* Josephus uses the term again and outlines the 'philosophical' beliefs and practices of the Pharisees (*A.J.* 18.2–3).

Although rabbis were not the direct successors of Pharisaic Judaism, they acted in a similarly Hellenised context in Roman Palestine.[7] Especially late antique rabbis who lived in the cities of Tiberias, Sepphoris, and Caesarea, that is, in urban contexts that are likely to have constituted the nodal points for the transmission, collection, and editing of rabbinic traditions,[8] would have been interested in presenting themselves to their coreligionists and perhaps also to non-Jews as scholars who were equal or superior to Graeco-Roman and Christian intellectuals whom they encountered in daily life. Rather than claiming a direct influence of late antique scholarly models on Palestinian rabbis, I argue that rabbis, philosophers, sophists, and church fathers represented variant versions of late antique intellectual culture that were partly overlapping and partly distinctive, geared at different constituencies. It should be noted, however, that rabbis do not fit the category of scholastic culture as it was understood from the time of the Renaissance onwards, with its image of the isolated scholar who contemplated metaphysical and philological issues in a merely theoretical way.[9] Their learning was closely linked to daily life and involved practical implementation as much as theoretical discussions with students and colleague-friends.[10]

The Social Contexts of Rabbinic and Graeco-Roman Intellectual Pursuits

Can late antique rabbis be defined as intellectuals? The answer to this question obviously depends on what we mean by 'intellectuals' and 'intellectualism' in antiquity. As Rita Copeland has already pointed out, '[a]ntiquity had no word for intellectuals... When historians of antiquity or the Middle Ages approach the question of intellectuals, it is always in terms of the formation of the social concept of modernity...'.[11] To avoid the imposition of a modern concept on ancient society, an alternative inductive and phenomenological approach needs to be chosen that starts with the ancient literary evidence itself. In the 'Introduction' to his study of *The Image of the Intellectual*

---

7 On the relationship between rabbis and Pharisees, see especially Cohen (2010); Schäfer (1991).
8 On the so-called urbanisation of the rabbinic movement in late antiquity, see Levine (1989) 25; Hezser (1997) 157–65; Lapin (1999); Lapin (2000).
9 On this image, see Cabezon (1998b) 2–3.
10 See also Swartz (1998) 106.
11 Copeland (2002) 40.

in Antiquity, Paul Zanker states that he uses the term 'simply as a convenient shorthand' for 'prophets, wise men, poets, philosophers, Sophists, and orators', who did not share 'a sense of group identity' but had a certain self-consciousness and were recognised by others as occupying 'a special position' within society.[12] Similarly, Kendra Eshleman includes Christian writers alongside philosophers and sophists in her study of 'Intellectuals in the Roman Empire'.[13] Interestingly, none of these scholars mentions rabbis in this connection. Rabbis and rabbinic scholarship are usually overlooked by classical scholars focusing on Graeco-Roman and Christian scholarship in the Roman and early Byzantine Empires.

Hayim Lapin views rabbis as a 'subaltern' elite, that is, as a group with a voice among the otherwise voiceless Palestinian Jewish subjects of Rome.[14] He writes: '... the rabbinic movement functioned in Palestine as an elite association of provincials inventing and fostering a local indigenous tradition'.[15] Whether they were considered elite by urban Jewish aristocrats and Graeco-Roman philosophers is questionable, though. As scholars, they lacked political and administrative offices; as Jewish scholars, they lacked Greek *paideia*, the type of education that distinguished members of the provincial elite. Greek *paideia* is often seen as the bond that linked members of the Graeco-Roman elite and created 'a sense of shared culture and identity'.[16] The so called 'triumph' of Christianity from the time of Constantine onwards is often linked to Christian leaders' foray into the upper strata of Roman-Byzantine society. Susanna Elm writes: 'The men who led the church before and after Constantine derived their authority from their social position, as members of the provincial elites. Members of that group, made homogeneous through shared education, or *paideia*, were groomed for positions of power and authority'.[17]

Although most rabbis would have lacked Greek *paideia* and did not belong to the upper strata of society, rabbinic literary sources present them as scholars with small circles of students who resembled Greek, Roman, and Christian scholars in a number of regards.[18] The model of the individual teacher surrounded by his students was also common in Graeco-Roman and Christian contexts, as many examples show. Willi Brown refers to 'the many philosophical study circles (διδασκαλεῖα) that gathered around guru-like guides'.[19] According to Peter Brown, '[s]mall study-circles were the powerhouses of the Christian culture of the second and third centuries'.[20] As far as pagan scholars were concerned, Plotinus (third century CE) claims that

12 Zanker (1995) 2.
13 Eshleman (2012).
14 Lapin (2012) 2–3.
15 Lapin (2012) 124.
16 Miles (2000) 48. See also Watts (2012) 468.
17 Elm (2012) 7.
18 On disciple circles, see Cohen (1999) 952–54.
19 Braun (2004) 45.
20 Brown (1988) 104.

he belonged to the small and exclusive circle of disciples surrounding the teacher Ammonius Saccas, a group that also included Longinus and Origen.[21] Porphyry refers to Plotinus' own disciple circle and Eunapius to Iamblichus and his disciples.[22] Similarly, Origen had a number of devoted students who remained loyal to him.[23] In late antique and early Byzantine times various degrees of formalisation and institutionalisation of schools existed. The School of Libanius seems to have been initially informal, with shifting teaching locations, but later he 'settled in the city hall on becoming Antioch's official sophist'.[24] Some of the large philosophical, sophistic, and rhetorical schools associated with the most famous scholars and orators were probably more formalised than rabbinic study sessions, but informal teaching also persisted alongside such institutions into the fourth and fifth centuries CE.[25]

Throughout the first to fifth centuries, rabbinic teaching, reflected in Palestinian rabbinic documents, was unofficial. It took place in small disciple circles that could meet in the rabbi's own house or at other places, rather than in fixed academies with hierarchical structures and established curricula.[26] Similarly unregulated and unofficial were meetings of rabbinic colleague-friends, where scholarly discussions took place. Although rabbinic texts occasionally mention study houses (*batei midrash*) at various locations, there is no evidence that these buildings—if they were buildings, for *bayit* can also mean 'room'—resembled the later academies that were established in Babylonia in stammaitic and geonic times.[27]

In an article published in the early 1980s Shaye Cohen has argued that the late antique patriarchs resembled the scholarchs of Hellenistic philosophical schools. His argument is almost exclusively based on texts transmitted in the Babylonian Talmud, however, and cannot be taken as evidence for the situation in Roman Palestine.[28] At the end of the article he admits: 'Perhaps then the parallels between patriarchs and scholarchs tell us more about the Hellenisation of Babylonian Jewry in the fourth and fifth centuries than about the Hellenisation of Palestinian Jewry in the second'.[29] In a different vein, Adam H. Becker has compared Babylonian rabbinic academies with East Syrian (Nestorian) Christian schools of the sixth century CE. These 'schools were no longer centred on charismatic teachers, but became actual *places* of learn-

---

21 See Fowden (1986) 129.

22 Heine (2010) 50.

23 Heine (2010) 50.

24 Cribiore (2007) 30.

25 I am grateful to Jan Stenger for alerting me to the continuity of informal teaching in study circles in late antique and early Byzantine times. He referred me to Eunapius of Sardis, *Lives of Sophists and Philosophers*; Marinus, *Life of Proclus*; Damascius, *Life of Isidor*; Gregory of Nyssa, *Life of Macrina*; Synesius of Cyrene, *Opuscules* as examples for such teaching.

26 See Hezser (2016) 18–22.

27 On the establishment of Babylonian academies, see especially Rubenstein (2003), 23 and 143. Rubenstein builds on Goodblatt (1975).

28 Cohen (1981).

29 Cohen (1981) 85.

ing. In the same way, the Babylonian rabbinic *yeshivot* were transformed from study circles centred on learned masters to formal institutions that remained intact after the death of individual masters'.[30]

How does Palestinian rabbinic scholarship fit into the social contexts of Greek, Roman, and Christian scholarship in late antiquity? One similarity was its foundation on an authoritative text as well as on oral traditions and interpretations that had come down to rabbis and informed rabbinic teachings. Whereas classical Greek *paideia* started with Homer, rabbinic study was ultimately based on the Torah.[31] The Torah constituted an ideational basis that informed rabbis' own decision-making. The goal of rabbinic education was to enable students to develop their own opinions in the spirit of the Torah, meant to provide guidance to their Jewish contemporaries in their daily lives. Similarly, Graeco-Roman philosophers developed their own systems based on earlier written texts and oral teachings.

The ultimate goal was to provide students with guidance in life. As Trevor Curnow has pointed out, 'ancient philosophy was about everyday life', helping people to make the right choices.[32] Like Graeco-Roman philosophy and Christian monastic teaching, rabbinic instruction was practically oriented. The rabbi, philosopher, and monk were believed to embody their teachings. As far as Cynic philosophy is concerned, Curnow has suggested that '[t]here was always an element of anti-intellectualism', with the Cynic himself serving as a 'living embodiment of Cynic teachings'.[33] Jan Stenger has emphasised the role of practice in the monk's progress toward virtue in Dorotheus of Gaza's teaching.[34] Similarly, rabbis discussed the relationship between study and practice and differed over what was more important.[35] A discussion of this issue appears in Sifre Deuteronomy 41:

> And it shall come to pass, if you listen carefully to my commandments' [*Deut 11:13*] – Why is it said? Since it is said, '[*Hear, oh Israel, the statutes and laws which I speak in your ears today.*] And you shall learn them and keep them to do them' [*Deut 5:1*], I might conclude that they [*the Israelites*] were not made liable to study [*the Torah*] before they became liable to practice [*maaseh, i. e., observe it*]...

Rabbis were interested in raising the importance of study, though. The discussion continues:

---

30 Becker (2010) 95. For a more detailed study of scholasticism in the Church of the East, see Becker (2006).

31 See Hezser (2016) 9–13.

32 Curnow (2006) 1.

33 Curnow (2006) 15.

34 Stenger (2017).

35 On the combination of study and practice and 'the complementary nature of Torah and good deeds', see also Schofer (2005) 69.

> 'And you shall learn them and keep them to do them' [*Deut 5:1*] – this tells that practice depends on study and study does not depend on practice. And so we find that a more severe penalty applies to [*the neglect of*] study than to [*the neglect of*] practice... (Sifre Deuteronomy 41).

A subsequent story about R. Tarfon, R. Aqiva, and R. Yose the Galilean, who are said to have discussed the issue, concludes with: 'All responded, saying: Great[er] is study, for study leads to practice' (Sifre Deuteronomy 41).

Another component of both rabbinic and Roman higher education was the element of socialising and identity creation. As Watts has emphasised, 'Although fundamentally literary in character, later Roman education did as much to socialise students as it did to familiarise them with the works of a classical canon'.[36] If a shared *paideia* created an elite consciousness, identification with particular philosophical and sophistic traditions led to distinctions among the intellectual elite and generated discussions and disputes, even within one and the same movement.[37] Similarly, a certain sage-consciousness developed among sages, which coexisted with divisions into small clusters of colleague-friends who agreed among themselves but might dispute other rabbis' opinions.[38]

Before looking more closely at rabbinic scholarship within the Graeco-Roman context, it is important to note that rabbis were probably less prone to use written material than Graeco-Roman scholars were, although much of philosophical training was also conducted orally.[39] In rabbinic study sessions and in scholarly discussions, even the Torah would usually have been cited from memory. As holy objects, Torah scrolls had to be handled with care. Written by specially trained scribes with ink on parchment, they would have been so expensive that they were usually communally owned (cf. y. Ned. 5:5, 39b: owned by 'the town'). Only very wealthy individuals would have been able to afford copies. While a few rabbis may have owned particular books of the Torah, direct access to the complete text would have been the exception rather than the rule. Only from the fifth century onwards did Palestinian synagogues have fixed Torah shrines.[40] Since rabbis were not leaders of synagogues in antiquity, their access to these scrolls may have been limited to liturgical contexts.

Throughout the Roman and early Byzantine period, memorisation of the text would have been a crucial part of both Jewish primary education and rabbinic study.[41] References in discussions would have been based on the memorised text, and this reliance on memory may have continued at the editorial and scribal stage in the creation of rabbinic documents. The situation was similar with tannaitic rab-

36 Watts (2012) 468.

37 On such disputes, see, for example, Barnes (1999) 40: between Peripatetics and Stoics; Xenakis (1969) 63, who talks about 'the endlessness of value disputes' between Stoics and non-Stoics.

38 On this issue, see Hezser (1997) 240 – 54.

39 For a more detailed discussion, see Hezser (2017).

40 Hachlili (1989) 3.

41 On ancient Jewish primary education, see Hezser (2001) 40 – 89; Hirshman (2009) is more concerned with rabbinic education.

binic traditions in amoraic times, which would have circulated orally.[42] Late antique rabbis were aware of their lack of books in contrast to Greeks, Roman, and Christians, but they interpreted this deficit as an advantage. A tradition attributed to R. Abin states: 'If I [i.e., God] had written down for you the larger part of my Torah, you would not be considered a stranger anymore [*cf. Hos 8:12*]'. Jews differed from Greeks and Romans because they owned the so-called Oral Torah, whereas others produced books (the term *sefer* is used here) and other pieces of writing (*diphthera*) (y. Peah 2:6, 17a par. y. Hagigah 1:8, 76d).

By contrast, Graeco-Roman philosophers who discussed matters orally seem to have also studied texts together. According to H. Gregory Snyder, the Stoics Persius and Cornutus 'spent many an hour together in literary pursuits, perhaps in the study of one or more of the 700 treatises of Chrysippus owned by Persius, but also, no doubt, in the study of Virgil and Homer'.[43] As members of the upper strata of society, they could own books themselves or had friends from whom they could borrow them. In late antique philosophical schools, students 'would listen to what amounted to line-by-line discussions of philosophical texts and their meaning'.[44] Despite their literate learning, Stoics could also be critical of bookishness: 'Although Seneca is a highly literate figure there is a palpable weariness and cynicism about bookish erudition that causes him to claim, 'wisdom is not to be found in letters' (Ep. 88.32)'.[45] Loveday Alexander has pointed to Galen's emphasis on 'the living voice' in teaching.[46] Although written books existed, teaching was mostly conducted orally, through questions and answers and discussions. Nevertheless, Snyder is probably also right when writing that 'philological pursuits were ... the order of the day in many Stoic teaching circles, and Stoics may be aptly described as bookish', a designation that does not apply in the same way to rabbis, for reasons already stated.[47]

Palestinian Rabbis and the Second Sophistic

Based on Philostratus' biographies of Greek sophists, scholars view the so-called Second Sophistic as a 'cultural phenomenon', a 'Greek renaissance' in the Roman Empire of the second and early third centuries CE.[48] With rabbinic Judaism the Sec-

42 For the relationship between orality and writing in rabbinic culture, see especially Jaffee (2001). He stresses that rabbis insisted upon 'the memorization and oral transmission of rabbinic *torah*' (154). Even texts that existed in writing would have been memorised and proclaimed orally in study sessions.
43 Snyder (2000), 41.
44 Watts (2006), 4.
45 Snyder (2000) 41.
46 Alexander (1990).
47 Snyder (2000) 42.
48 See the book title and introduction in Anderson (1993). For a discussion of the term 'Second Sophistic', see Bowersock (1985) 655–62.

ond Sophistic seems to have shared an interest in using cultural traditions of the past to create particular Jewish and Hellenic identities. Rabbis and sophists emphasised the importance of a particular type of education, whether Torah study or *paideia*. Knowledge of the literary tradition of the past was relevant for understanding and properly functioning within the present. Just as Greek imperial literature 're-appropriates the Greek traditions of the Classical age of the fifth and fourth centuries BC', rabbis claimed that Torah study was the necessary prerequisite for leading a Jewish life after the destruction of the Second Temple in 70 CE.[49]

As Borg has pointed out, as a Greek revival movement the Second Sophistic 'is not a purely literary phenomenon, but a value system and mode of thought', that also included (non-sophist) authors and texts not directly mentioned by Philostratus.[50] Bowersock refers to Artemidorus, who was not a member of the movement but 'an explicit witness to the anti-Roman sentiments that others have claimed to detect as implicit in the writings of the sophists themselves'.[51] Sophists were not only orators. Artemidorus sometimes refers to them as teachers: 'They are linked with διδάσκαλοι and with παιδευταί'.[52] Some sophists lived in or visited Palestine and its vicinity,[53] and Geiger has suggested that Greek-speaking rabbis, who lived in cities such as Caesarea, may have been among their audience.[54]

The sophistic movement encompassed many different types of scholarly expertise and expressed itself in a variety of literary forms.[55] What seems to be clear, though, is that sophists sought to uphold the value of the Greek tradition at the time of Roman imperialism.[56] According to Simon Swain, Greek writers emphasised 'Hellenic culture and speech as the emblems of civilization' in their confrontation with Rome. He emphasises that '[t]he Greek elite's cultural power at this time is at stark contrast with its lack of overall political independence and effective political representation'.[57] The rabbinic analogy is quite obvious. Rabbis, who also lacked official political power, endeavoured to generate and preserve a distinctly Jewish tradition in Hebrew and Aramaic at a time when 'indigenous' Jewish culture was threatened by the Romanisation of Palestine and the Near East.

Anthony J. Saldarini has already pointed out that Palestinian rabbis of the first centuries lived in a cultural context that was marked by the Second Sophistic. Especially rabbis such as the patriarch R. Yehudah ha-Nasi, the alleged editor of the Mish-

49 On the Second Sophistic, see Borg (2004) 1.

50 Borg (2004) 2. On the extent of the sophistic movement beyond the sophists mentioned by Philostratus, see also Eshleman (2012) 125–48.

51 Bowersock (2004) 59.

52 Bowersock (2004) 60, with reference to Artemidorus 2.12; 2.45; 4.13.

53 For a discussion of the evidence, see Geiger (1994).

54 Geiger (1994) 230.

55 See Whitmarsh (2005) 2; Bowersock (1969).

56 See Bowie (1974).

57 Swain (1996), 2.

nah, who is associated with Greek learning, would have been aware of this movement. Saldarini writes: 'Like other intellectuals during the Second Sophistic, Rabbi used his learning and influence for social leadership. His formulation of the Mishnaic code took place in the context of the intellectual efforts of the Second Sophistic, when Greco-Roman authors created digests, compendia, and codes, and Christian authorities began the canonization of the New Testament'.[58] Saldarini does not investigate this context any further, though. Whereas Saldarini saw an analogy between the compilation of the Mishnah and Greek compilation projects, Yair Furstenberg has suggested the need to study rabbinic scriptural exegesis (*midrash*) in the context of the rhetorical education of the Second Sophistic.[59] This comparative approach can ultimately be traced down to Saul Lieberman and Henry Fischel, but has not been continued much in recent rabbinic scholarship.[60]

In the following discussion I shall focus on two aspects: the issue of collecting, preserving, and compiling earlier traditions and the significance of the dispute form in rabbinic texts. In these and other regards the editors of Palestinian rabbinic documents can be compared with Greek scholars of the Roman imperial period.

The Compilation of Earlier Traditions

The compilers of the Mishnah created a collection of rabbinic traditions that constituted a Jewish alternative to what they considered 'Greek wisdom'. This compilation of tannaitic traditions would serve as the basis for further rabbinic study in the amoraic period of the third and fourth centuries CE and eventually constitute the basis of the Palestinian Talmud, edited in the late fourth or early fifth century CE. Although the compilation of the Mishnah has traditionally been associated with the first patriarch R. Yehudah ha-Nasi, there is no reliable historical information about the compilers, their methods, and goals. As the most well-connected rabbi, R. Yehudah ha-Nasi probably played a role in the compilation, but he must have collaborated with other rabbis of his time, some of whom disagreed with his opinions.[61]

The collecting and writing down of mostly oral traditions, their reformulation, combination, and arrangement into discursive units, tractates, and orders would have been a long and complex process that may have taken decades to accomplish. All we can say is that amoraic rabbis of the third and fourth centuries were aware of a collection of earlier rabbinic traditions called Mishnah, which they considered authoritative and on which they based their own argumentation. Whether and to what extent this collection circulated and was accessed in written form remains uncertain

58 Saldarini (1994) 15.

59 Furstenberg (2012).

60 Lieberman (1962); Lieberman (1965); Fischel (1977).

61 On the question of R. Yehudah ha-Nasi's role in editing the Mishnah, see Halivni (2013) 103: 'The Editor of the Mishnah and His Role in Editing'.

and may have varied from one rabbi and locale to the next.[62] It needs to be emphasised, however, that the Mishnah did not constitute a comprehensive and binding law code. Christine Hayes has already pointed to 'its many contradictions and unresolved disputes, its lack of comprehensiveness, and its sometimes obscure organization and arrangement'.[63] It is therefore more likely that the Mishnah was meant to preserve earlier legal traditions that could serve as the basis of further rabbinic study by later generations of scholars. The goals of the compilers would therefore have been antiquarian, in the sense of preserving traditions of the past, as well as scholastic, making them available for study purposes.

Can we compare the compilation of the Mishnah, which was preserved in Hebrew in the early third century CE, to Greek compilations of approximately the same time period? As already mentioned, the Mishnah is not a law code, so legal compilations are not the (only) appropriate genres of comparison.[64] With regard to its significance as the foundation document of rabbinic Judaism the Mishnah is comparable to the New Testament, which became the basis of Graeco-Roman Christianity and patristic scholarship. Yet its content, form, and structure are quite dissimilar. Vernon K. Robbins has suggested that the structuring of a Mishnaic chapter can be understood on the basis of rhetorical categories outlined in Hermogenes' *Progymnasmata.*[65] He points to 'the distinctive manner in which this text [*the Mishnah*] uses and reconfigures conventional patterns of argumentation in Mediterranean society and culture'.[66] In the Mishnah, as in the *Progymnasmata*, rhetorical elaboration of a statement and rule can take the form of anecdote or *chreia.*[67] Both the Mishnah and the *Progymnasmata* were school literature used in educational frameworks. Yet the contents and goals of rabbinic and rhetorical education were different.[68]

The compilation of the Mishnah (and the later Palestinian Talmud) is probably best understood in the context of what Jason König and Gregg Woolf call encyclopaedism: 'the ways in which a series of different authors (...) made use of a range of shared rhetorical and compilatory techniques to create knowledge-ordering works of different kinds, work that often claimed some comprehensive and definitive status'.[69] The resulting texts would be part of 'an encyclopaedic spectrum, with different texts drawing on shared encyclopaedic markers to different degrees and for very different purposes'.[70] The focus is not on the identification of a particular liter-

---

62 On the question of the Mishnah's publication, see Hezser (2002).
63 Hayes (2017) 111.
64 Usually, the Mishnah has been compared to law codes, see already Neusner (2002) 122.
65 See Robbins (2002).
66 Robbins (2002) 201.
67 See Robbins (2002) 203.
68 See also Lightstone (2002) 65: 'Being a rabbinic master seems quite different from, or askew of, being a Greco-Roman rhetor ...'.
69 König/Woolf (2013a) 1.
70 König/Woolf (2013a) 1.

ary genre but on compilatory processes that were employed in order to preserve and transmit knowledge of the past to future generations.

Among the 'spectrum of shared techniques', mentioned by König and Woolf, the Mishnah exhibits the following:[71]

1) Excerption and recombination: The Mishnah (and even more so rabbinic Midrashim and the Talmud) uses words, phrases, and (parts of) verses from the Bible that have been taken out of their original literary contexts and are used to elicit, support, contradict, or illustrate rabbinic arguments. Within one unit of discourse excerpts from different biblical books can be combined and juxtaposed.[72]
2) Selection of pre-existing knowledge: The compilers of the Mishnah collected and selected traditions associated with tannaitic rabbis who lived between 70 and 200 CE. Most of these traditions would have been transmitted in oral form from one generation of rabbinic scholars to the next. As Teresa Morgan has observed on the basis of Greek wisdom collections, all of these works are 'digests of selected material' that 'covered the essentials of morality'.[73]
3) Ordering of the material: The received material, once (re)formulated and written down by scribes, was used to create thematic units of discourse that became parts of tractates and orders. It is important to note that the editors' work was not limited to the mere ordering of the material. The editors created argumentative units on the basis of the received material that were entirely new and partly reflected their viewpoints.
4) Decisions about structure: At some point within the editorial process the editors of the Mishnah must have decided to create tractates (sg. *massechet*) and to gather the tractates in six orders (sg. *seder*). Nevertheless, within individual tractates, a clear thematic focus is not consistently applied and the reasons that led to specific arrangements of discursive units usually remain unclear.
5) Ideal of comprehensiveness: We may assume that the editors aimed at providing a comprehensive collection of earlier rabbinic traditions. In reality, however, comprehensiveness would have been impossible to achieve. Continuous transmission depended on a continuous line of students and students' students over many generations. If the line of succession was interrupted, the teachings of earlier scholars could easily become lost. In addition, the 'final' editors who were active at the beginning of the third century would have primarily or exclusively collected material from their colleague-friends and from those rabbis that lived close to their own locations. These circumstances would have had an impact on the representativeness of the material that found its way into the compilation.

---

71 For a discussion of these techniques, see König/Woolf (2013a) 6–13. The authors never mention the Mishnah or other rabbinic compilations in this context. Their considerations are based on Graeco-Roman and Christian texts.

72 For a detailed study of the use of the Bible in the Mishnah, see Samely (2002).

73 Morgan (2013) 113.

6) Accessibility: Obviously, one of the main purposes of the compilation would have been to make earlier rabbinic knowledge accessible to future generations. This does not mean, however, that individual rabbis' views on particular subjects could be found easily or that this was even intended by the editors of the Mishnah. On the contrary, one may argue that the editors arranged the material in such a way that a continuous discussion of the halakhic topics rather than reliance on one particular rabbi's view was intended.[74]
7) Anonymity: The editors of the Mishnah do not identify themselves but remain anonymous. Some scholars have suggested that they stand behind sages' collective views that are sometimes juxtaposed with the variant views of named rabbinic authorities.[75] Some Graeco-Roman collections have named authors (e. g., Plutarch's *Sympotic Questions*, Valerius Maximus' *Memorable Deeds and Sayings*, Gellius' *Attic Nights*), while the editors of others (e. g., the various types of *florilegia*, *gnomologia*, and *epitomae*) did not identify themselves.
8) Polyphony: The Mishnah—and other rabbinic compilations as well—presents many different rabbinic opinions side by side, creating what in music is called a texture of simultaneous but independent voices or melodies.[76] König and Woolf see this phenomenon as a general feature of encyclopaedic works: 'The encyclopaedic text, almost by definition, contains many voices brought into close juxtaposition with each other'.[77]

The Mishnah shares these and other aspects with other ancient encyclopaedic works. It would therefore be very useful to make more detailed comparisons between the compilation and editing strategies of rabbinic, Graeco-Roman, and Christian collections of late antiquity, an undertaking that cannot be accomplished in the framework of this chapter.[78] König and Woolf have referred to 'recurring similarities of political context ... even across very different cultures'.[79] The political context of the Romanisation of Palestine may have motivated rabbis to preserve 'indigenous' Jewish knowledge and to make it accessible to Jewish scholars of later generations. Similarly, the Greek writers and compilers of the Second Sophistic had the preservation of Greek philosophical, literary, rhetorical, and legal knowledge in mind. Perhaps the Mishnah and Greek compilations can be seen as expressions of resistance to Roman au-

74 König/Woolf (2013b) 45, have made a similar observation with regard to Graeco-Roman compilations: they are not necessarily 'easily usable as storage devices from which readers might rapidly retrieve particular items of information'.

75 See, for example, Shemeh (2009) 66, who continues the traditional identification of the editors with R. Yehudah ha-Nasi, however, and views sages' opinions as authoritative.

76 For the Merriam-Webster online dictionary's definition, see https://www.merriam-webster.com/dictionary/polyphony.

77 König/Woolf (2013a) 13.

78 I hope to be able to conduct a larger research project on this topic in the near future.

79 König/Woolf (2013a) 11.

tocracy and imperialism. Even more so than Greek authors, whom the Romans may have studied themselves, the rabbinic compilers of the Mishnah created an alternative body of scholarship in Hebrew that was meant for internal Jewish (and even more narrowly rabbinic) use only. It contained the legal traditions of the 'native' Palestinian rabbinic intelligentsia, who would have competed with Roman legal experts as far as the adjudication of civil law was concerned.[80] The phenomenon that the issues discussed in the Mishnah are all-encompassing, addressing many aspects of daily life, indicates that rabbis were interested in controlling their fellow-Jews behaviour toward each other as much as toward non-Jews, including Greeks and Romans, whom they encountered in their cities and market places.

The Use of the Dispute Form

Probably the most striking feature of rabbinic texts, from the Mishnah to the Talmud, is the use of the dispute form. Variant rabbinic opinions are juxtaposed, often with no or little harmonisation between them. The passage may begin with a question, followed by two or three different answers, a short narrative that may support one of the answers or provide yet another option, and perhaps a more general statement at the end. This structure is found at the very beginning of Mishnah tractate Berakhot, where the recitation of the evening *Shema* (prayer) is discussed:

> [A] From what time onwards do they recite the *Shema* in the evening?
> [B] From the hour when the priests entered [*their houses*] to eat from their *terumah* [*heave offering*] until the first wake,[81] the words of R. Eliezer.
> [C] And sages say: Until midnight.[82]
> [D] Rabban Gamliel says: Until the pillar of dawn arises.[83]
> [E] A story [*maaseh*] that his sons came from a festive banquet. They said to him: We have not [*yet*] recited the *Shema*. He said to them: If the pillar of dawn has not yet arisen, you are [*still*] obliged to recite. And not in this case alone, but wherever sages have said 'until midnight', their rule is until the pillar of dawn arises...[84]
> [F] If so, why did sages say: 'until midnight'?
> [G] In order to keep a human being far away from transgression (m. Ber. 1:1).

While rabbis agreed that the *Shema* should be recited in the evening, they disagreed over the time limits within which the recitation of the prayer could be seen as the fulfilment of one's obligation. Underlying this discussion is Deuteronomy 6:7,

80 On judicial diversity in Roman Palestine see Harries (2010).
81 Meaning the end of the first third of the night, when people would usually lie down to sleep, as explained by Albeck (1988) 13 *ad loc.*
82 Which may be later than the first wake.
83 I.e., even later.
84 It is unclear whether this last sentence is still part of R. Gamliel's statement or an anonymous editorial addition.

where Israelites are instructed to 'talk about' God's words 'when you lie down [at night] and when you get up [in the morning]'. Based on this formulation, rabbis considered the twice daily recitation of the *Shema* obligatory for Jewish men. In order to actually practice the commandment, a more concrete time scale was required, though. The biblical text's general formulation gave rise to three different interpretations of how the divine command could actually be practiced.

The dispute form is part and parcel of rabbinic biblical interpretation. It is the outcome of the biblical base text's, and probably any text's, multiple meanings. To explicitly state a variety of meanings, and to continue to generate new meanings, increases the potency and relevance of the base text while, at the same time, preventing homogeneity in observance. To narrow down the base text's meanings would lead to a more unified ritual practice but also close the study and discussion of the text. As the example from the Mishnah shows, rabbis vacillate between the two possibilities and sometimes try to harmonise between diverse opinions. Although one fulfils one's obligation when reciting the *Shema* until 'the pillar of dawn arises', sages have said 'until midnight', that is, they set a narrower margin to prevent their fellow-Jews from (accidentally) transgressing the rule (cf. [F] – [G]). Multiple interpretations were generated because the Torah was a text that was open for interpretation. Harmonisations were sometimes necessary because the Torah was believed to be a sacred text that expressed God's will, which Jews were supposed to followed in daily life, so concrete instruction on how to practice the Torah were necessary.[85]

The prevalence of dialogue and debate in the Mishnah and in rabbinic Judaism in general seems to reveal something fundamental about the rabbinic notion of truth and the ways in which it can be approximated. For rabbis, truth was equivalent to God's will as it was expressed in the Torah. In order to find that truth, the Torah had to be studied. Yet the Torah was a text that could be interpreted in many different ways. Ultimately, humans could only approximate the truth but never know it perfectly. Therefore, different rabbinic interpretations can stand side by side and be considered equally valid. This notion is expressed in the (Vienna manuscript version of the) Tosefta, where contradictions between the houses of Hillel and Shammai are pointed out. The Tosefta legitimises contradictions by stating:

> Scripture says: 'Words... the words... these are the words...' [*Deut 1:1*]: All these words have been given by a single shepherd, one God made them, one Provider gave them, the Lord of all deeds, Blessed Be He, has spoken it. So you, open many chambers in your heart and bring into it the words of the house of Shammai and the words of the house of Hillel, the words of those who declare unclean and the words of those who declare clean (t. Sotah 7:12).

---

85 In his chapter on 'Disputes in Rabbinic Judaism', Neusner (2004) 229 recognises but also plays down the significance of the dispute form. He states: 'In the Halakhic documents, the Mishnah, Tosefta, Yerushalmi, and Bavli, Rabbinic sages ubiquitously record disagreements on matters of law. But disputes reinforce the unity of the law at its fundamental levels. Conflicts between authorities underscore the prevailing consensus about fundamental truth'.

Here disagreement and polyphony are traced back to the divine itself. As Boyarin has already emphasised, in this statement 'disagreement itself, or at any rate the appearance of disagreement to humans, is exemplary of the divine mind'.[86] The ideal is 'polynoia, the many-mindedness, as it were, of God'.[87] Boyarin is wrong, however, in assuming that 'such declarations are to be found only in the latest layers of classical rabbinic literature'.[88] He quotes the Babylonian Talmudic version of the statement only.[89]

Simon Goldhill has contrasted the prevalence of dialogue in Graeco-Roman and rabbinic culture with the apparent lack of it in ancient Christian literary sources. He associates dialogue with democracy, dissent, and persuasion.[90] Both the staged debate and dialogue as a literary form allegedly originated in classical Athens.[91] In dialogue as in democracy, the participating parties act on the basis of the principle of equality. In the Roman Empire, the dialogue form was 'adopted' by writers such as Cicero; 'sympotic dialogues' obtained an important role within Roman culture.[92] Christian writers, on the other hand, resisted dialogue, claiming 'a single truth' that was revealed once for all, propagating 'a strict hierarchical world picture'.[93]

Seth Schwartz has already argued that 'the rabbinic adoption of the sympotic style went far beyond the ritual of the seder'.[94] Rabbinic texts use the settings of the *symposium* or *convivium* for their own purposes to present themselves as socialising and discussing matters of mutual interest with their colleague-friends.[95] In this they resembled the Stoics, who had already adapted the form to express their own philosophical concerns. In rabbinic literature philosophical discussions are replaced by dialogues on Torah issues. The symposium was the ideal setting for intellectual dialogues among like-minded scholars. Within this framework, disagreement was possible. At the same time, the image of rabbinic socialising at meals and on other occasions created a framework of convivality and agreement that constituted the basis for disputes to take place. As Shaye Cohen has already pointed out, rabbis 'agreed to disagree'.[96] Disagreements on the basis of shared concerns constituted the essence of rabbinic identity and advanced rabbinic Torah scholarship.[97]

---

86 Boyarin (2004) 162.
87 Boyarin (2004) 162.
88 Boyarin (2004) 162.
89 On the talmudic version in b. Eruvin 13b (with a parallel in y. Yevamot 1:6, 3b), see Kraemer (1996) 65–67.
90 Goldhill (2008) 1.
91 Goldhill (2008) 2.
92 Goldhill (2008) 5. For examinations of these texts, see Schofield (2008); König (2008).
93 Goldhill (2008) 7.
94 Schwartz (2008) 208.
95 See the discussion in Hezser (2017a) 112–20.
96 Cohen (1984) 46.
97 Boyarin (2008) 220, argues that 'a loss of faith in rational discourse' and 'a mood of disappointment with rhetoric and argumentation' was the consequence of the early Byzantine church councils

Conclusion

Palestinian rabbinic literature presents rabbis as a particularly Jewish, 'indigenous' type of intellectuals, who functioned alongside other types of intellectuals in the scholastic milieu of the Roman and early Byzantine Near East. While Hellenised Jews of the upper strata of society and Greek intellectuals may not have perceived rabbis as members of the elite, at least some of their fellow Jews would have acknowledged their self-perception as a scholarly elite and followed individual rabbis' advice on how to lead a Torah-observant lifestyle. With their emphasis on Torah study and observance rabbis offered a local Jewish alternative to the empire-wide focus on Greek *paideia* that became the distinguishing mark of the educated upper strata of society, including Hellenised urban Jews.

Like some philosophical, sophistic, and monastic teachers, rabbis attracted more or less small groups of disciples and held informal meetings with small clusters of like-minded colleague-friends in private and semi-private settings, without fixed curricula and meeting places.[98] The teaching and discussion was probably less text-based than philosophical study, with Torah portions, rabbinic interpretations, and legal opinions quoted orally, on the basis of memorisation, rather than in the presence of book scrolls or written notes. The focus was on the Torah, the first and mostly legal part of the Hebrew Bible, with the goal of adapting Torah law to contemporary situations, issues, and circumstances. Discussions were probably conducted in Aramaic, the local language of the Near East. Since the Torah was written in Hebrew, Hebrew maintained an important religious role in rabbinic Judaism.[99] It is also the language in which the Mishnah, Tosefta, and tannaitic Midrashim are written. Judith Hauptman has suggested that the editors may be 'promoting Hebrew to a population that is abandoning it for Aramaic and Greek'.[100] The choice of language, together with the focus on the Torah rather than Homer, suggests that rabbis intentionally created an alternative Jewish form of scholarship that could hold up to Greek *paideia* within their local Near Eastern context.[101]

The creation of this particular type of Jewish scholarship has to be understood within the context of the loss of the Second Temple in 70 CE and Roman imperialism

---

and had an impact on the way in which disputes are presented in the later Babylonian Talmud. The editors of the Babylonian Talmud often harmonise between disagreements, in contrast to the Palestinian Talmuds, where disputes tend to remain unresolved.

98 For a detailed discussion of the question of the rabbinic movement's institutionalisation, see Hezser (1997) 185–227.

99 On the religious significance of Hebrew and Aramaic as the local language of the Near East, see Schwartz (1995).

100 Hauptman (2005) 207.

101 There has been a lot of speculation about the question why rabbinic Judaism did not spread beyond its Near Middle Eastern environment in antiquity and the early Middle Ages. On this issue, see Mendels/Edrei (2011).

that led to the Romanisation and later the Christianisation of the Jewish Land of Israel in the first five centuries CE. The rabbinic focus on Torah study and the creation of a vast body of rabbinic scholarship would have functioned as an alternative to what rabbis called 'Greek wisdom', that is, Greek *paideia* and Hellenistic culture, which Roman intellectuals and artists were eager to adopt for their own purposes.[102]

Despite functioning as a local alternative to Greek scholarship, rabbinic scholarship evinced many analogies to Graeco-Roman philosophical, rhetorical, and legal study. Not only are some of the settings similar—the combination of seated sessions and *perambulatio*; the mostly oral teaching and discussion in small circles of teachers and students or among like-minded colleague-friends—but there is a shared notion that theoretical wisdom should find expression in lifestyle choices, that the 'wise man' should serve as a model that could be observed and imitated.

As far as scholarly pursuits and editorial procedures are concerned, tannaitic scholarship can be studied within the context of and compared with the so-called Second Sophistic movement of the first three centuries CE. Like scholars of the Second Sophistic, rabbis were concerned with the preservation of earlier rabbinic traditions. The Mishnah was compiled in the early third century to make the (mostly orally transmitted) traditions of the preceding centuries available to future generations of scholars who would enter the process of constant reinterpretation, adaptation, and searching for solutions to new problems. One particular literary form that served these purposes was the dispute form which is all pervasive in the Mishnah, Tosefta, and Talmud. The use of the dispute form indicates that, for rabbis, the truth (i.e., the meaning of the Torah as God's will) could only be approximated, that it might actually exist in many forms (i.e., rabbinic views) within the framework of a rabbinically perceived monotheism.

102 Scholarship on the Roman adaptation of Greek and Hellenistic culture is vast. Here we can only point to a few such studies: Gruen (1990); Goldhill (2001).

Bibliography

Adams (2018): Adams, S. A. 'Where's Rome? A Possible Roman Parallel to the Translation of the Septuagint in the *Letter of Aristeas*'. In G. Xeravits and G. Schmidt (eds.), *Figures Who Shape Scriptures, Scriptures That Shape Figures: Essays in Honour of Benjamin G. Wright III*. DCLS 40. Berlin: de Gruyter, 197–210.

Adler (1938): Adler, A. 'Urbano fra Belluno's Graeske Haandskrifter'. *Nordisk Tidskrift för Bok- och Biblioteksväsen* 25, 73–80.

Ahl (1971): Ahl, Frederick. 'Lucan's *De incendio urbis*, *Epistulae ex Campania* and Nero's Ban'. *TAPhA* 102, 1–27.

Aitken (2011): Aitken, James K. 'The Literary Attainment of the Translator of Greek Sirach'. In J.-S. Rey and J. Joosten (eds.), *The Texts and Versions of the Book of Ben Sira: Transmission and Interpretation*. *JSJSup* 150. Leiden: Brill, 95–126.

Aitken (2014): Aitken, James K. *No Stone Unturned: Greek Inscriptions and Septuagint Vocabulary*. CSHB 5. Winona Lake, IN: Eisenbrauns.

Aitken (2016): Aitken, James K. 'The Septuagint and Egyptian Translation Methods'. In W. Kraus, M.N. van der Meer, and M. Meiser (eds.), *XV Congress of the International Organization for Septuagint and Cognate Studies: Munich, 2013*. *SBLSCS* 64. Atlanta: SBL Press, 269–93.

Albeck (1988): Albeck, Hanoch. *The Six Orders of the Mishnah: Seder Zera'im* [Hebrew]. Tel Aviv: Dvir.

Alberti (1972–2000): Alberti, G. B. *Thucydidis Historiae, I-III*. Rome: Istituto Poligrafico dello Stato.

Albl (1999): Albl, M. C. *'And Scripture Cannot Be Broken': The Form and Function of the Early Christian Testimonia Collections*. *NovTSup*, 96. Leiden: Brill.

Albl (2016): Albl, M. C. 'The *Testimonia* Hypothesis and Composite Citations'. In S.A. Adams and S.M. Ehorn (eds.), *Jewish, Graeco-Roman, and Early Christian Uses*, vol. 1 of *Composite Citations in Antiquity*. *LNTS* 525. London: Bloomsbury T&T Clark, 2016, 182–202.

Alexander (1990): Alexander, Loveday. 'The Living Voice: Scepticism Towards the Written Word in Early Christian and in Graeco-Roman Texts'. *Journal for the Study of the Old Testament Supplement Series* 87, 221–47.

Alpers (1975): Alpers, Klaus. 'Review of *Contributions to a History of Alphabetisation in Antiquity and the Middle Ages*, by Lloyd W. Daly'. *Gnomon* 47, 113–18.

Alpers (1981): Alpers, Klaus. *Das attizistische Lexikon des Oros: Untersuchung und kritische Ausgabe der Fragmente*. Berlin: de Gruyter.

Alpers (2004): Alpers, Klaus. 'Die griechischen Orthographien aus Spätantike und byzantinischer Zeit. Anmerkungen zu einer Publikation'. *Byzantinische Zeitschrift* 97, 1–50.

Ammirati (2010): Ammirati, Serena. 'Per una storia del libro latino antico: osservazioni paleografiche, bibliologiche e codicologiche sui manoscritti latini di argomento legale dalle origini alla tarda antichità'. *Journal of Juristic Papyrology* 40, 55–110.

Ammirati (2013): Ammirati, Serena. 'The use of wooden tablets in the ancient Graeco-Roman world and the birth of the book in codex form: some remarks'. *Scripta: An International Journal of Codicology and Paleography* 6, 9–15.

Ammirati (2015a): Ammirati, Serena. *Sul libro latino antico. Ricerche bibliologiche e paleografiche*. Pisa-Rome: Fabrizio Serra Editore.

Ammirati (2015b): Ammirati, Serena. 'Leggere Cicerone in Egitto: osservazioni paleografiche (e filologiche)'. In Paolo De Paolis (ed.), *Dai papiri al XX secolo. L'eternità di Cicerone. Atti del VI Simposio ciceroniano. Arpino, 9 maggio 2014* (XXXIV Certamen Ciceronianum Arpinas). Cassino: Cassino University Press, 13–15.

https://doi.org/10.1515/9783110660982-012

Ammirati (2017): Ammirati, Serena. 'Il paratesto nei manoscritti giuridici di origine romana'. *Segno e Testo* 15, 247 – 60.
Ammirati (forthcoming): Ammirati, Serena. 'Il codice veronese delle Institutiones di Gaio. Paleografia e codicologia'. In Ulrike Babusiaux and D. Mantovani (eds.), *Le istituzioni di Gaio: avventure di un best-seller: Trasmissione, uso e trasformazione del testo.* Pavia: Pavia University Press.
Anderson (1993): Anderson, Graham. *The Second Sophistic: A Cultural Phenomenon in the Roman Empire.* London: Routledge.
Armstrong (2008): Armstrong, R. 'Classical Translations of the Classics: The Dynamics of Literary Tradition in Retranslating Epic Poetry'. In A. Lianeri and V. Zajko (eds.), *Translation and the Classic: Identity as Change in the History of Culture.* Oxford: Oxford University Press, 169 – 202.
von Arnim (1895): von Arnim, Hans. 'Apollonios von Tyros [94]'. *RE* 2.1, 146.
Asmis (2001): Asmis, E. 'Basic Education in Epicureanism'. In Y.L. Too (ed.), *Education in Greek and Roman Antiquity.* Leiden: Brill, 209 – 39.
Aufrère *et al.* (2014): Aufrère, S. H., P. S. Alexander, and Z. Pleše, editors. *On the Fringes of Commentary: Metatextuality in Ancient Near Easter and Ancient Mediterranean Cultures.* OLA 232. Leuven: Peeters.
Ax (1990): Ax, Wolfram. 'Aristophanes von Byzanz als Analogist: zu Fragment 374 Slater (= Varro, *de lingua Latina* 9,12)'. *Glotta* 68, 4 – 18.
Badoud (2002): Badoud, Nathan. 'La table claudienne de Lyon au XVIe siècle'. *Cahiers du Centre Gustave Glotz* 13, 169 – 95.
Baldwin (1983): Baldwin, Barry. *Suetonius.* Amsterdam: Hakkert.
Barclay (1996): Barclay, J. M. G. *Jews in the Mediterranean Diaspora: From Alexander to Trajan (323 BCE-117 CE).* Edinburgh: T&T Clark.
Bardon (1952): Bardon, Henry. *La littérature latine inconnue.* Vol. 1: *L'époque républicaine.* Paris: Klincksieck.
Barker (1820): Barker, E. H. Ἀρκαδίου Περὶ Τόνων: *e codicibus parisinis.* Leipzig: Gehrard Fleischer.
Barnes (1999): Barnes, Jonathan. 'Aristotle and Stoic Logic'. In Katerina Ierodiakonou (ed.), *Topics in Stoic Philosophy.* Oxford: Oxford University Press, 23 – 53.
Baron (2013): Baron, Christopher A. *Timaeus of Tauromenion and Hellenistic historiography.* Cambridge: Cambridge University Press.
Barr (1979): Barr, J. 'The Typology of Literalism in Ancient Biblical Translations'. Mitteilungen des Septuaginta-Unternehmens, 15. Gottingen: Vandenhoeck & Ruprecht.
Bassnett (2014): Bassnett, S. *Translation Studies.* Fourth edition. London: Routledge.
Bayer (2002): Bayer, Karl. *Suetons Vergilvita: Versuch einer Rekonstruktion.* Tübingen: Gunter Narr.
Beaulieu (2006): Beaulieu, Paul-Alain. 'Berossus on Late Babylonian History'. *Oriental Studies,* 116 – 49.
Becker (2006): Becker, Adam H. *Fear of God and the Beginnings of Wisdom. The School of Nisibis and the Development of Scholastic Culture in Late Antique Mesopotamia.* Philadelphia, PA: University of Pennsylvania Press.
Becker (2010): Becker, Adam H. 'The Comparative Study of 'Scholasticism' in Late Antique Mesopotamia: Rabbis and East Syrians'. *AJS Review* 34, 91 – 113.
Bernstein (2005): Bernstein, M. J. '"Rewritten Bible": A Generic Category Which Has Outlived Its Usefulness?'. *Textus* 22, 169 – 96.
Biville (2002): Biville, F. 'The Graeco-Romans and Graeco-Latin: A Terminological Framework for Cases of Bilingualism'. In J.N. Adams, M. Janse, and S. Swain (eds.), *Bilingualism in Ancient Society: Language Contact and the Written Text.* Oxford: Oxford University Press, 77 – 102.

Blank (1994): Blank, David L. 'Analogy, Anomaly and Apollonius Dyscolus'. In Stephen Everson (ed.), *Language*. Cambridge: Cambridge University Press, 149–165.
Blank (1998): Blank, David L. *Sextus Empiricus: Against the grammarians*. Oxford: Clarendon Press.
Bloomer (2011): Bloomer, W. M. *The School of Rome: Latin Studies and the Origins of Liberal Education*. Berkeley, CA: University of California Press.
Blum (1991): Blum, Rudolf. *Kallimachos: The Alexandrian Library and the Origins of Bibliography*. Translated by Hans Wellisch. Madison, Wis.: University of Wisconsin Press.
Bollansée (1999): Bollansée, Jan. *Hermippos of Smyrna and his Biographical Writings: A Reappraisal*. Leuven: Peeters.
Boffo (1995): Boffo, Laura. 'Ancora unavolta sugli 'archivi' nel mondo greco: conservazione e 'publicazione' epigrafica'. *Athenaeum* 83, 91–130.
Bonner (1977): Bonner, S. F. *Education in the Roman World: From the Elder Cato to the Younger Pliny*. Berkeley: University of California.
Borchard (2012): Borchard, F. 'The LXX Myth and Rise of Textual Fixity'. *JSJ* 43, 1–21.
Borg (2004): Borg, Barbara E. 'Introduction'. In Barbara E. Borg (ed.), *Paideia: The World of the Second Sophistic*. Berlin: de Gruyter, 1–10.
Bossi (2000): Bossi, F. 'Sui Περιεργοπένητες di Diogeniano'. *Eikasmós* 11, 267–68.
Boudon-Millot *et al.* (2010): Boudon-Millot, V. and J. Jouanna with A. Pietrobelli eds. *Galien, Oeuvres Tome IV: Ne pas se chagriner*. Paris: Les Belles Lettres.
Boussac/Invernizzi (1996): Boussac, Marie-Françoise and Antonio Invernizzi, eds. *Archives et sceaux du monde hellénistique, Bulletin de correspondance hellénique supplement 29*. Athens: École française d'Athènes.
Bowersock (1969): Bowersock, Glenn Warren. *Greek Sophists in the Roman Empire* Oxford: Clarendon Press.
Bowersock (1985): Bowersock, Glenn Warren. 'Philostratus and the Second Sophistic'. In P.E. Easterling and Bernard M.W. Knox (eds.), *Cambridge History of Classical Literature*, vol. 1: *Greek Literature*. Cambridge: Cambridge University Press, 655–62.
Bowersock (2004): Bowersock, Glenn Warren. 'Artemidorus and the Second Sophistic'. In Barbara E. Borg (ed.), *Paideia. The World of the Second Sophistic*. Berlin: de Gruyter, 53–64.
Bowie (1974): Bowie, Ewen. 'The Greeks and Their Past in the Second Sophistic'. In Moses I. Finley (ed.), *Studies in Ancient Society*. London: Routledge, 166–209.
Bowie (1978): Bowie, Ewen. 'Apollonius of Tyana: Tradition and Reality'. *ANRW* 2.16.2, 1652–99.
Bowie (2013): Bowie, Ewen. 'Libraries for the Caesars'. In Jason König, Katerina Oikonomopoulou and Greg Woolf (eds.), *Ancient Libraries*. Cambridge: Cambridge University Press, 237–60.
Boyarin (2004): Boyarin, Daniel. *Border Lines: The Partition of Judaeo-Christianity*. Philadelphia, PA: University of Pennsylvania Press.
Boyarin (2008): Boyarin, Daniel. 'Dialectic and divination in the Talmud'. In Simon Goldhill (ed.), *The End of Dialogue in Antiquity*. Cambridge: Cambridge University Press, 217–41.
Braun (2004): Braun, Willi. 'The Schooling of a Galilean Jesus Association'. In Ronald Dean Cameron and Merill P. Miller (eds.), *Redescribing Christian Origins*. Leiden: Brill, 43–65.
Brink (1963): Brink, Charles. *Horace on Poetry: Prolegomena to the Literary Epistles*. Cambridge: Cambridge University Press.
Brock (1972): Brock, S. P. 'The Phenomenon of the Septuagint'. *OtSt* 17, 11–36.
Brock (1977): Brock, S. P. 'Greek into Syriac and Syriac into Greek'. *Journal of the Syriac Academy* 3, 1–16.
Brock (1979): Brock, S. P. 'Aspects of Translation Technique in Antiquity'. *GRBS* 20, 69–87.
Brock (1992): Brock, S. P. 'To Revise or Not to Revise: Attitudes to Jewish Biblical Translation'. In G.J. Brooke and B. Lindars (eds.), *Septuagint, Scrolls and Cognate Writings*. *SBLSCS* 33. Atlanta: Scholars Press, 301–28.

Broggiato (2000): Broggiato, M. 'Athenaeus, Crates and Attic Glosses: a problem of attribution'. In David Braund and John Wilkins (eds.), *Athenaeus and his World: Reading Greek Culture in the Roman Empire*. Exeter: Exeter University Press, 364 – 71.

Brooke (2010): Brooke, G. J. 'Genre Theory, Rewritten Bible, and Pesher'. *DSD* 17, 332 – 57.

Brooke (2017): Brooke, G. J. 'Aspects of Education in the Sectarian Scrolls from the Qumran Caves'. In G.J. Brooke and R. Smithuis (eds.), *Jewish Education from Antiquity to the Middle Ages: Studies in Honour of Philip S. Alexander. AJEC* 100. Leiden: Brill, 11 – 42.

Brown (1988): Brown, Peter. *The Body and Society: Men, Women, and Sexual Renunciation in Early Christianity*. New York, NY: Columbia University Press.

Brugnoli (1968): Brugnoli, Giorgio. *Studi Suetoniani*. Lecce: Milella.

Brugnoli (1984): Brugnoli, Giorgio. *Foca: Vita di Virgilio: introduzione, testo, traduzione e commentato*. Pisa: ETS.

Brugnoli/Stock (1997): Brugnoli, Giorgio, and Stock, Fabio. *Vitae Vergilianae antiquae*. Rome: Typis Officinae Polygraphicae.

Bravo (2007): Bravo, Benedetto. 'Antiquarianism and History'. In John Marincola (ed.), *A companion to Greek and Roman historiography*. London: Blackwell, 515 – 27.

De Breucker (2010): De Breucker, Geert. 'Berossos of Babylon (680)'. In Ian Worthington (ed.), *Brill's New Jacoby*. Leiden: Brill.

De Breucker (2013): De Breucker, Geert. 'Berossos: His life and his work'. In Johannes Haubold, Giovanni B. Lanfranchi, Robert Rollinger, and John Steele (eds.), *The World of Berossos. Proceedings of the 4th International Colloquium on the Ancient Near East between classical and ancient oriental traditions, Hartfield College, Durham 7th-9th July 2010*. Wiesbaden: Harrassowitz, 15 – 28.

Burrell (2009): Burrell, B. 'Reading, Hearing and Looking at Ephesos'. In W.A. Johnson and H.N. Parker (eds.), *Ancient Literacies. The Culture of Reading in Greece and Rome*. Oxford: Oxford University Press, 69 – 95.

Butler (2008 – 2009): Butler, Shane. 'Cicero's capita'. *Litterae Caelestes* 3, 9 – 48.

Cabezon (1998a): Cabezon, José Ignacio, edited. *Scholasticism: Cross-Cultural and Comparative Perspectives*. Albany, NY: SUNY Press.

Cabezon (1998b): Cabezon, José Ignacio. 'Introduction'. In José Ignacio Cabezon (ed.), *Scholasticism: Cross-Cultural and Comparative Perspectives*. Albany, NY: SUNY Press, 1 – 17.

Canevaro/Harris (2012): Canevaro, Mirko and Edward M. Harris. 'The Documents in Andocides' On the Mysteries'. *Classical Quarterly* 62, 98 – 129.

Capasso (2003): Capasso, Mario. *Il ritorno di Cornelio Gallo: il papiro di Qasr Ibrîm venticinque anni dopo. Con un contributo di Paolo Radiciotti*. Napoli: Graus.

Capizzi (1955): Capizzi, Antonio. *Protagora: le testimonianze e i frammenti*. Florence: G. C. Sansoni.

Carandini/Carafa (2017): Carandini, A. and P. Carafa, editors. *The Atlas of Ancient Rome*. Princeton and Oxford: Princeton University Press.

Carawan (2007): Carawan, Edwin. 'Krateros the Macedonian (342)'. In Ian Worthington (ed.), *Brill's New Jacoby*. Leiden: Brill.

Cavallo (2002): Cavallo, Guglielmo. 'Un'aggiunta al 'decalogo' di Giorgio Pasquali'. In *Dalla parte del libro. Storie di trasmissione dei classici*. Urbino: QuattroVenti, 25 – 29.

Cerati/Locke (1755): Cerati, G. and J. Locke. 'Copy of a letter from a learned Gentleman of Naples, dated February 25, 1755, concerning the books and antient writings dug out of the ruins of an edifice near the site of the old city of Herculaneum; to Monsignor Cerati, of Pisa, F. R. S. Sent to Mr. Baker, FRS and by him communicated; with a translation by John Locke, Esq FRS'. *Philosophical Transactions of the Royal Society* 48, XXIII, 112 – 15.

Chandler (1881): Chandler, H. W. *A practical introduction to Greek Accentuation*. Oxford: Oxford University Press.

Christophilopoulou (1979): Christophilopoulou, Aikaterini P. 'Η καταχωρισις δικαιοπρακτικων εγγραφων εις τα δημοσια αρχεια'. *Νομικα Επιγραφικα* 2, 9–69.
Clancier (2009): Clancier, Philippe. *Les bibliothèques en Babylonie dans la deuxième moitié du 1er millénaire av. J.-C.* Münster: Ugarit-Verlag.
Claridge (2007): Claridge, A. 'Hadrian's lost temple of Trajan'. *Journal of Roman Archaeology* 20, 54–94.
Claridge (2013): Claridge, A. 'Hadrian's Succession and the Monuments of Trajan'. In T. Opper (ed.), *Hadrian: Art, Politics and Economy*. London: British Museum, 5–18.
Clarke (1973): Clarke, Martin. 'The date of the Ciris'. *CPh* 68, 119–21.
Clarysse (2010): Clarysse, W. 'Bilingual Papyrological Archives'. In A. Papaconstantinou (ed.), *The Multilingual Experience in Egypt, from the Ptolemies to the Abbasids*. Farnham: Ashgate, 47–72.
Coarelli (2010): Coarelli, F. 'Substructio et Tabularium'. *Papers of the British School at Rome* 78, 107–32.
Coates-Stevens (2006): Coates-Stevens, R. 'Un *pistrinum* tardo-repubblicano a Porta Maggiore'. *Rendiconti della Pontificia Accademia Romana di Archeologia* 78, 473–98.
Cohen (1981): Cohen, Shaye J.D. 'Patriarchs and Scholarchs'. *Proceedings of the American Academy for Jewish Research* 48, 57–87.
Cohen (1984): Cohen, Shaye J.D. 'The Significance of Yavneh: Pharisees, Rabbis, and the End of Jewish Sectarianism'. In *The Significance of Yavneh and Other Essays in Jewish Hellenism*. Tübingen: Mohr Siebeck, 44–70. Originally published in: *Hebrew Union College Annual* 55 (1984): 27–53.
Cohen (1999): Cohen, Shaye J.D. 'The Rabbi in Second-Century Jewish Society'. In William Horbury, W.D. Davis, and John Sturdy (eds.), *The Cambridge History of Judaism*, vol. 3: *The Early Roman Period*. Cambridge: Cambridge University Press, 922–90.
Cohn (1895): Cohn, L. 'Arkadios 5'. In *RE* 2.1, 1153–56.
Cooke/Tredennick (1938): Cooke, H. P., and H. Tredennick, editors. *Aristotle: Categories, On Interpretation, Prior Analytics*. LCL 325. Cambridge, Mass: Harvard University Press.
Cooper (2007): Cooper, C., editor. *Politics of Orality*. MneSup 280. OLAC 6. Leiden: Brill.
Copeland (1991): Copeland, Rita. *Rhetoric, Hermeneutics, and Translation in the Middle Ages: Academic Traditions and Vernacular Texts*. Cambridge: Cambridge University Press.
Copeland (2002): Copeland, Rita. 'Pre-Modern Intellectual Biography'. In Helen Small (ed.), *The Public Intellectual*. Oxford: Blackwell Publishers, 40–61.
Coqueugniot (2012): Coqueugniot, Gaëlle. 'Le chreophylakeion et l'agora d'Europos-Doura: bilan des recherches, 2004–2008'. In Pierre Leriche, Gaëlle Coqueugniot, and Ségolène de Pontbriand (eds.), *Europos-Doura Varia I*. Beirut: Presses de l'ifpo, 93–110.
Coqueugniot (2013): Coqueugniot, Gaëlle. *Archives et bibliothèques dans le monde grec. Édifices et organisation*. Oxford: BAR.
Corcoran/Salway (2012): Corcoran, Simon and Benet Salway, 'Fragmenta Londiniensia Anteiustiniana: Preliminary Observations'. *Roman Legal Tradition* 8, 63–83.
Cornell *et al.* (2013): Cornell, Tim, et al. *The Fragments of the Roman Historians*. 3 Vols. Oxford: Oxford University Press.
Courcelle (1968): Courcelle, Pierre. *Late Latin Writers and their Greek Sources*. Translated by Harry Wedeck. Cambridge: Harvard University Press.
Courtney (1993): Courtney, Edward. *The Fragmentary Latin Poets*. Oxford: Clarendon Press.
Cribiore (1996): Cribiore, Raffaella. *Writing, Teachers, and Students in Graeco-Roman Egypt*. ASP 36. Atlanta: Scholars Press.
Cribiore (2001): Cribiore, Raffaella. *Gymnastics of the Mind: Greek Education in Hellenistic and Roman Egypt*. Princeton: Princeton University Press.

Cribiore (2007): Cribiore, Raffaella. *The School of Libanius in Late Antique Antioch*. Princeton, NJ: Princeton University Press.

De Crom (2011): De Crom, D. 'Translation and Directionality in the Hebrew-Greek Tradition'. In S. McElduff and E. Sciarrino (eds.), *Complication the History of Western Translation: The Ancient Mediterranean in Perspective*. Manchester: St. Jerome, 77–78.

Cugusi (2007): Cugusi, Paolo. *Per un nuovo corpus dei Carmina Latina Epigraphica: materiali e discussioni*. Roma: Bardi.

Cugusi (2016): Cugusi, Paolo. *Versi su pietra: studi sui Carmina latina epigraphica: metodologia, problemi, tematiche, rapporti con gli auctores, aspetti filologici e linguistici, edizioni di testi: quaranta anni di ricerche*. Faenza: Fratelli Lega editori.

Curnow (2006): Curnow, Trevor. *Ancient Philosophy and Everyday Life*. Newcastle: Cambridge Scholars Press.

D'Anna (1956): D'Anna, Giovanni. 'Sulla vita suetoniana di Terenzio'. *RIL* 89, 31–46.

D'Ors (1986): D'Ors, Alvaro. *La ley Flavia Municipal (Texto y comentario)*. Rome: Pontificia Universitas Lateranensis.

Dahlmann (1963): Dahlmann, Hellfried. *Studien zu Varro De poetis*. Mainz: F. Steiner.

Dahlmann (1973): Dahlmann, Hellfried. 'Varroniana'. *ANRW* 1.3, 2–25.

Davila (2005): Davila, J. R. '(How) Can We Tell if a Greek Apocryphon or Pseudepigraphon has been Translated from Hebrew or Aramaic?'. *JSP* 15, 3–61.

Decorte (2015): Decorte, Robrecht. 'Publishing Laws: An Investigation of Layout and Epigraphic Conventions in Roman Statutes'. *Zeitschrift für Papyrologie und Epigraphik* 195, 243–54.

Del Corso (2009): Del Corso, L. 'L'Insegnamento alla luce delle testimonianze iconografiche'. In H. Huggonard-Roche (ed.), *L'enseignement supérieur dans les mondes antiques et médiévaux. Textes et traditions 16*. Paris: Editions VRIN, 307–31.

Dickey (2007): Dickey, Eleanor. *Ancient Greek Scholarship: A Guide to Finding, Reading, and Understanding Scholia, Commentaries, Lexica, and Grammatical Treatises, from Their Beginnings to the Byzantine Period*. Oxford: Oxford University Press.

Dickey (2010): Dickey, Eleanor. 'The creation of Latin teaching materials in antiquity: a re-interpretation of P.Sorb. inv. 2069'. *Zeitschrift für Papyrologie und Epigraphik* 175, 188–208.

Dickey (2012–2015): Dickey, Eleanor. *The Colloquia of the Hermeneumata Pseudodositheana*. 2 vols. CCTC. Cambridge: Cambridge University Press.

Dickey (2014): Dickey, Eleanor. 'A Catalogue of Works Attributed to the Grammarian Herodian', *Classical Philology* 109, 325–45.

Dickey (2016): Dickey, Eleanor. *Learning Latin the ancient way: Latin textbooks from the ancient world*. Cambridge: Cambridge University Press.

Diels/Kranz (1952): Diels, Hermann, and Walther Kranz. *Die Fragmente der Vorsokratiker* (6th edn). Berlin: Weidmann.

Dillery (1999): Dillery, John. 'The first Egyptian narrative history: Manetho and Greek Historiography'. *Zeitschrift für Papyrologie und Epigraphik* 127, 93–116.

Dillery (2013): Dillery, John. 'Manetho'. In Tim Whitmarsh and Stuart Thomson (eds.), *The Romance between Greece and the East*. Cambridge: Cambridge University Press, 38–58.

Dindorf (1823): Dindorf, C. W. *Grammatici Graeci, I*. Leipzig: Libraria Kuehniana.

Dines (2004): Dines, J. M. *The Septuagint*. London: T&T Clark.

Dix/Houston (2006): Dix, Keith and George Houston. 'Public Libraries in the City of Rome: From the Augustan Age to the Time of Diocletian'. *MEFRA* 118. 671–717.

Dixon (1997): Dixon, R. M.W. *The Rise and Fall of Languages*. Cambridge: Cambridge University Press.

Dorandi (1999): Dorandi, Tiziano. *Antigone de Caryste*. Paris: Les Belles Lettres.

Dorandi (2007): Dorandi, Tiziano. *Nell'officina dei classici: come lavoravano gli autori antichi*. Rome: Carocci.
Dingel (1973): Dingel, Joachim. 'Bruchstück einer römischen Komödie auf einem Hamburger Papyrus (Afranius?)'. *Zeitschrift für Papyrologie und Epigraphik* 10, 29–44.
Dover (1968): Dover, Kenneth J. *Aristophanes: Clouds*. Oxford: Clarendon Press.
Dreyer (1975): Dreyer, O. 'Luseis'. *KP*, 832–33.
Dudley (1967): Dudley, D. R. *Urbs Roma: a source book of classical texts on the city and its monuments*. London: Phaidon.
Dyck (1993): Dyck, A. R. 'Aelius Herodian: Recent Studies and Prospects for Future Research'. In *ANRW* II 34.1, 772–94.
École nationale (1991): École nationale des Chartes, *Dictionnaire des archives. De l'archivage aux systèmes d'information*. Paris: AFNOR.
Eckstein (2010): Eckstein, Arthur M. 'Polybius, 'the treaty of Philinus' and Roman accusations against Carthage'. *The Classical Quarterly* 60, 406–26.
Edmondson (2014): Edmondson, Jonathan. 'Inscribing Roman texts: Officinae, layout, and carving techniques'. In Christer Bruun and Jonathan Edmondson (eds.), *The Oxford Handbook of Latin Epigraphy*. Oxford: Oxford University Press, 111–30.
Edwards (2012): Edwards, M. J. 'Religion in the Age of Marcus Aurelius'. In M. Van Ackeren (ed.), *A companion to Marcus Aurelius*. Chichester: Wiley-Blackwell, 200–16.
Egenolff (1881): Egenolff, P. 'In Herodianum technicum critica III'. *RhM* 36, 490–505.
Egenolff (1887): Egenolff, P. *Die orthoepischen Stücke der byzantinischen Litteratur*. Leipzig: Teubner.
Egenolff (1900): Egenolff, P. '"Zu Lentz" Herodian I'. *Philologus* 59, 238–55.
Egenolff (1902): Egenolff, P. '"Zu Lentz" Herodian II", *Philologus* 61, 77–132, 540–76.
Egenolff (1903): Egenolff, P. '"Zu Lentz" Herodian III'. *Philologus* 62, 39–63.
El-Abbadi (2004): El-Abbadi, M. 'The Alexandria library in history'. In Anthony Hirst and Michael Silk (eds.), *Alexandria, Real and Imagined*. London: Routledge, 167–84.
Elm (2012): Elm, Susanna. *Sons of Hellenism, Fathers of the Church: Emperor Julian, Gregory of Nazianzus, and the Vision of Rome*. Berkeley: University of California Press.
Erbse (1969–1988): Erbse, Hartmut. *Scholia graeca in Homeri Iliadem*. Berlin: De Gruyter.
Erskine (1995): Erskine, A. 'Culture and power in Ptolemaic Egypt: The Museum and Library of Alexandria'. *Greece & Rome* 42, 38–48.
Eshleman (2012): Eshleman, Kendra. *The Social World of Intellectuals in the Roman Empire: Sophists, Philosophers, and Christians*. Cambridge: Cambridge University Press.
Esposito (2009): Esposito, E. 'Fragments of Greek lexicography in the papyri'. *Trends in Classics* 1, 255–97.
Esposito (2011): Esposito, E. 'Review of *From Alexandria to Babylon. Near Eastern Languages and Hellenistic Erudition in the Oxyrhynchus Glossary (POxy 1802 + 4812)*, by Francesca Schironi'. *Eikasmós* 22, 524–29.
Feeney (2016): Feeney, D. *Beyond Greek: The Beginnings of Latin Literature*. Cambridge, MA: Harvard University Press.
Fernández Marcos (2000): Fernández Marcos, N. *The Septuagint in Context: Introduction to the Greek Version of the Bible*. Translated by W.G.E. Watson. Leiden: Brill.
Fernández Pomar (1966): Fernández Pomar, J. M. 'La Colección de Uceda y los manuscritos griegos de Constantino Lascaris'. *Emerita* 34, 211–88.
Fewster (2002): Fewster, P. 'Bilingualism in Roman Egypt'. In J.N. Adams, M. Janse, and S. Swain (eds.), *Bilingualism in Ancient Society: Language Contact and the Written Text*. Oxford: Oxford University Press, 220–46.

Fioretti (2012): Fioretti, Paolo. 'Ordine del testo, ordine dei testi. Strategie distintive nell'Occidente latino tra lettura e scrittura'. In *Scrivere e leggere nell'alto medioevo. Spoleto, 28 aprile – 4 maggio 2011*, 525–51. Spoleto: Centro Italiano di Studi sull'Alto Medioevo.

Fioretti (2016): Fioretti, Paolo. 'Percorsi di autori latini tra libro e testo. Contesti di produzione e di ricezione in epoca antica'. *Segno e testo* 14, 1–38.

Fischel (1977): Fischel, Henry A. *Essays in Greco-Roman and Related Talmudic Literature*. New York, NY: Ktav.

Fowden (1986): Fowden, Garth. *The Egyptian Hermes: A Historical Approach to the Late Pagan Mind*. Cambridge: Cambridge University Press.

Fraenkel (1950): Fraenkel, Eduard. *Aeschylus Agamemnon*. Oxford: Clarendon Press.

Fraenkel (1957): Fraenkel, Eduard. *Horace*. Oxford: Clarendon Press.

Frampton (2019): Frampton, S. *Empire of Letters*. Oxford: Oxford University Press.

Fraser (1972): Fraser, P. M. *Ptolemaic Alexandria*. 3 vols. Oxford: Clarendon Press.

Fraser (2009): Fraser, P. M. *Greek Ethnic Terminology*. Oxford: Oxford University Press.

Fraser *et al.* (1987-): Fraser, P. M., E. Matthews, et al. (eds.), *A lexicon of Greek personal names*. Oxford: Oxford University Press.

Freudenthal (1875): Freudenthal, J. *Alexander Polyhistor und die von ihm erhaltenen Reste jüdischer und samaritanischer Geschichtswerke, Hellenistische Studien I-II*. Breslau: Skutsch.

Fressura (2017): Fressura, Marco. *Vergilius Latinograecus. Corpus dei manoscritti bilingui dell'Eneide. Parte prima (1–8)*. Pisa-Rome: Fabrizio Serra editore.

Fressura/Mantovani (2018): Fressura, Marco and Dario Mantovani, 'P.Vindob. L 59 + 92. Frammenti delle Institutiones di Elio Marciano'. *Athenaeum* 106, 619–690.

Funaioli (1907): Funaioli, Gino. *Grammaticae Romanae fragmenta*. Leipzig: Teubner.

Furstenberg (2012): Furstenberg, Yair. 'The Agon With Moses and Homer: Rabbinic Midrash and the Second Sophistic'. In Maren R. Niehoff (ed.), *Homer and the Bible in the Eyes of Ancient Interpreters*. Leiden: Brill, 299–328.

Galland (1882): Galland, C. *De Arcadii qui fertur libro de accentibus*. Strasbourg: Carolus J. Truebner.

Gallazzi (2012): Gallazzi, Claudio. 'Umm-el-Breigât (Tebtynis): campagna di scavo dell'anno 2012'. *Istituto Lombardo – Accademia di Scienze e Lettere – Rendiconti di Lettere* 146, 87–110.

Gamillscheg *et al.* (1981-): Gamillscheg, E., D. Harlfinger, and D. Hunger, *Repertorium der Griechischen Kopisten*. Vienna: Austrian Academy of Sciences Press.

Gartner (2002): Gartner, H. A. 'Zetema'. *DNP* 12, 778–79.

Gartner (2010): Gayraud, M. 'L'apprentissage du grec et du latin dans l'Empire romain d'après un manuscrit de la Bibliothèque Universitaire de Montpellier'. *Bulletin de l'Académie des Sciences et Lettres de Montpellier* 41, 35–44.

Geiger (1985): Geiger, Joseph. *Nepos and Political Biography*. Wiesbaden: Steiner.

Geiger (1994): Geiger, Joseph. 'Notes on the Second Sophistic in Palestine'. *Illinois Classical Studies* 19, 221–30.

Georgoudi (1988): Georgoudi, Stella. 'Manières d'archivage et archives de cités'. In Marcel Detienne (ed.), *Les savoirs de l'écriture. En Grèce ancienne*. Lille: Presses universitaires de Lille, 221–47.

Gigante (1995): Gigante, M. *Philodemus in Italy. The Books from Herculaneum*. Translated by Dirk Obbink. Ann Arbor: University of Michigan Press.

Gignac (1981): Gignac, Francis T. *A grammar of the Greek papyri of the Roman and Byzantine periods* II. Milan: Istituto editoriale cisalpino La goliardica.

Goldhill (2001): Goldhill, Simon, edited. *Being Greek Under Rome: Cultural Identity, the Second Sophistic and the Development of Empire*. Cambridge: Cambridge University Press.

Goldhill (2008): Goldhill, Simon. 'Introduction: Why don't Christians do dialogue?'. In Simon Goldhill (ed.), *The End of Dialogue in Antiquity*. Cambridge: Cambridge University Press, 1–11.

Goldschmidt (2015): Goldschmidt, Nora. 'Terence: A Guide to Selected Sources'. In *Living Poets*. Durham.
Gomoll (1935): Gomoll, Heinz. 'Suetons bibliotheksgeschichtliche Nachrichten'. *Zentralblatt für Bibliothekswesen* 52, 381–88.
González (1986): González, Julián. 'The Lex Irnitana: A New Copy of the Flavian Municipal Law'. *Journal of Roman Studies* 76, 147–243.
Goodblatt (1975): Goodblatt, David. *Rabbinic Instruction in Sasanian Babylonia*. Leiden: Brill.
Goodman (2016): Goodman, P. 'Clusters of artisans in the Roman city'. In A. Wilson and M. Flohr (eds.), *Urban Craftsmen and Traders in the Roman World*. Oxford: Oxford University Press, 301–33.
Goodwin (1894): Goodwin, William W. *A Greek Grammar*. Boston: Ginn & Company.
Goodyear (1965): Goodyear, Francis. *Incerti Auctoris Aetna*. Cambridge: Cambridge University Press.
Goodyear (1984): Goodyear, Francis. 'The 'Aetna': Thoughts, Antecedents, and Style'. *ANRW* 2.32.1, 344–64.
Greer (1926): Greer, Russel. 'Non-Suetonian Passages in the Life of Vergil Formerly Ascribed to Donatus'. *TPAPhS* 57, 107–15.
Grenfell/Hunt (1900): Grenfell, Bernard and Arthur S. Hunt, editors. *Fayum Towns and Their Papyri*. London: Egypt Exploration Society.
Griffin (1984): Griffin, Jasper. 'Augustus and the Poets. *Caesar qui cogere posset*'. In Fergus Millar and Erich Segal (eds.), *Caesar Augustus: Seven Aspects*. Oxford: Clarendon Press, 189–218.
Gruen (1990): Gruen, Erich S. *Studies in Greek Culture and Roman Policy*. Berkeley: University of California Press.
Gudeman (1927): Gudeman, A. 'Λύσεις'. *PW* 13/2, 2511–29.
Gunderson (2009): Gunderson, E. *Nox Philologiae. Aulus Gellius and the Fantasy of the Roman Library*. Madison: University of Wisconsin Press.
Haake (2013): Haake, Matthias. 'Psephismata in the Biographies of Philosophers'. In Peter Liddel and Polly Low (eds.), *Inscriptions and Their Uses in Greek and Latin Literature*. Oxford: Oxford University Press, 79–124.
Hachlili (1989): Hachlili, Rachel. 'The State of Ancient Synagogue Studies'. In Rachel Hachlili (ed.), *Ancient Synagogues in Israel. Third-Seventh Century C.E.* Oxford: British Archaeological Review, 1–5.
Hahm (1992): Hahm, David. 'Diogenes Laertius VII: On the Stoics'. *ANRW* 2.36.2, 4076–182.
Halivni (2013): Halivni, David Weiss. *The Formation of the Babylonian Talmud*. Introduced, Translated, and Annotated by Jeffrey L. Rubenstein. Oxford: Oxford University Press.
Halton (1999): Halton, Thomas. *Saint Jerome: On Illustrious Men*. The Fathers of the Church 100. Washington D.C.: Catholic University Press of America.
Hansen (2016): Hansen, Mogens H. 'Is Teisamenos' Decree (Andoc. 1.83–84) a Genuine Document?'. *Greek, Roman, and Byzantine Studies* 56, 34–48.
Harding (2001): Harding, Phillip. *Androtion and the Atthis*. London: Clarendon.
Harding (2008): Harding, Phillip. *The Story of Athens: The Fragments of the Local Chronicles of Attika*. London: Routledge.
Harries (2010): Harries, Jill. 'Courts and the Judicial System'. In Catherine Hezser (ed.), *The Oxford Handbook of Jewish Daily Life in Roman Palestine*. Oxford: Oxford University Press, 85–101.
Hartnett (2017): Hartnett, J. *The Roman Street: Urban Life and Society in Pompeii, Herculaneum, and Rome*. Cambridge: Cambridge University Press.
Hartog (2017): Hartog, P. B. *Pesher and Hypomenma: A Comparison of Two Commentary Traditions from the Hellenistic-Roman Period*. *STDJ* 121. Leiden: Brill.

Hatzilambrou (2007): Hatzilambrou, R. '*P. Oxy.* XVIII 2192 revisited'. Appendix to Dirk Obbink. 'Readers and intellectuals' in *Oxyrhynchus: a City and its Texts.* Edited by Alan Bowman, pages 282–86. London: Egypt Exploration Society.

Hatzimichali (2013a): Hatzimichali, Myrto. 'Encyclopedism in the Alexandrian Library'. In Jason König and Greg Woolf (eds.), *Enclyclopaedism from Antiquity to the Renaissance.* Cambridge: Cambridge University Press, 64–83.

Hatzimichali (2013b): Hatzimichali, Myrto. 'Ashes to Ashes? The library of Alexandria after 48 BC'. In Jason König, Katerina Oikonomopoulou and Greg Woolf (eds.), *Ancient Libraries.* Cambridge: Cambridge University Press, 167–82.

Hatzimichali (2016): Hatzimichali, Myrto. 'Andronicus of Rhodes and the construction of the Aristotelian Corpus'. In Andrea Falcon (ed.), *Brill's Companion to the Reception of Aristotle in Antiquity.* Leiden: Brill, 81–100.

Hauptman (2005): Hauptman, Judith. *Rereading the Mishnah: A New Approach to Ancient Jewish Texts.* Tübingen: Mohr Siebeck.

Hayes (2017): Hayes, Christine. 'Law in Classical Rabbinic Judaism'. In Christine Hayes (ed.), *The Cambridge Companion to Judaism and Law.* Cambridge: Cambridge University Press, 76–127.

Heine (2010): Heine, Ronald E. *Origen: Scholarship in the Service of the Church.* Oxford: Oxford University Press.

Heitsch (1964): Heitsch, Ernst. *Die griechischen Dichterfragmente der römischen Kaiserzeit* II. Göttingen: Vandenhoeck & Ruprecht.

Helm (1929): Helm, Rudolf. *Hieronymus' Zusätze in Eusebius' Chronik und ihr Wert für die Literaturgeschichte.* Philologus Supplementband 21.2. Leipzig: Dieterich'sche Verlagsbuchhandlung.

Hengel (1974): Hengel, M. *Judaism and Hellenism: Studies in their Encounter in Palestine during the Early Hellenistic Period.* Translated by J. Bowden. Minneapolis: Fortress Press.

Hermans (2002): Hermans, T. 'Paradoxes and Aporias in Translation and Translation Studies'. In Alessandra Riccardi (ed.), *Translation Studies: Perspectives on an Emerging Discipline.* Cambridge, Cambridge University Press, 10–23.

Hezser (1997): Hezser, Catherine. *The Social Structure of the Rabbinic Movement in Roman Palestine.* Tübingen: Mohr Siebeck.

Hezser (2001): Hezser, Catherine. *Jewish Literacy in Roman Palestine. TSAJ* 81. Tübingen: Mohr Siebeck.

Hezser (2002): Hezser, Catherine. 'The Mishnah and Ancient Book Production'. In Jacob Neusner and Alan J. Avery-Peck (eds.), *The Mishnah in Contemporary Perspective,* vol. 1. Leiden: Brill, 167–92.

Hezser (2016): Hezser, Catherine. 'The Torah Versus Homer: Jewish and Greco-Roman Education in Late Roman Palestine'. In Matthew Ryan Hauge and Andrew W. Pitts (eds.), *Ancient Education and Early Christianity.* LNTS 533. London: Bloomsbury, 5–24.

Hezser (2017a): Hezser, Catherine. *Rabbinic Body Language: Non-Verbal Communication in Palestinian Rabbinic Literature of Late Antiquity.* Leiden: Brill.

Hezser (2017b): Hezser, Catherine. 'Bookish Circles? The Use of Written Texts in Rabbinic Oral Culture'. *Temas Medievales* 25, 63–81.

Hezser (2018): Hezser, Catherine. 'The Creation of the Talmud and the Apophthegmata Patrum as Monuments to the Rabbinic and Monastic Movements in Early Byzantine Times'. *Jewish Studies Quarterly* 25, 368–93.

Hilgard (1901): Hilgard, A. *Scholia in Dionysii Thracis Artem grammaticam* (*Grammatici Graeci I.III*). Leipzig: Teubner.

Hirshman (2009): Hirshman, Marc. *The Stabilization of Rabbinic Culture, 100 C.E. – 350 C.E. Texts On Education and Their Late Antique Context.* Oxford: Oxford University Press.

Holford-Strevens (2003): Holford-Strevens, Leofrank. *Aulus Gellius: An Antonine Scholar and his Achievement.* Oxford: Oxford University Press.

Holford-Strevens/Vardi (2004): Holford-Strevens, L. and A. Vardi, editors. *The Worlds of Aulus Gellius.* Oxford: Oxford University Press.

Holleran (2012): Holleran, C. *Shopping in Ancient Rome: The Retail Trade in the Late Republic and the Principate.* Oxford: Oxford University Press.

Holwerda (1991): Holwerda, D. *Scholia in Vespas, Pacem, Aves et Lysistratam (Scholia in Aristophanem II.3: Scholia vetera et recentiora in Aristophanis Aves).* Groningen: Egbert Forsten.

Honigman (2003): Honigman, Sylvie. *The Septuagint and Homeric Scholarship in Alexandria: A Study in the Narrative of the 'Letter of Aristeas'.* London: Routledge.

Honigman (2007): Honigman, Sylvie. 'The Narrative Function of the King and the Library in the *Letter of Aristeas*'. In T. Rajak *et al. Jewish Perspectives on Hellenistic Rulers.* Berkeley: University of California Press, 128–46.

Horsfall (1989): Horsfall, Nicholas. *Cornelius Nepos: A Selection, including the Lives of Cato and Atticus.* Oxford: Clarendon Press.

Houston (2007): Houston, George W. 'Grenfell, Hunt, Breccia and the book collections of Oxyrhynchus'. *Greek, Roman and Byzantine Studies* 47, 327–59.

Houston (2014): Houston, George W. *Inside Roman Libraries: Book Collections and Their Management in Antiquity.* Studies in the history of Greece and Rome. Chapel Hill: University of North Carolina Press.

Hudson (1980): Hudson, R. A. *Sociolinguistics.* CTL. Cambridge: Cambridge University Press.

Hunger (1967): Hunger, H. 'Palimpsest-Fragmente aus Herodians *Καθολικὴ Προσῳδία*, Buch 5–7'. *Jahrbuch der Österreichischen Byzantinischen Gesellschaft* 16, 1–33.

Hunt (1927): Hunt, Arthur S., editor *The Oxyrhynchus Papyri*, XVII, *Nos. 2065–2156.* London: Egypt Exploration Society.

Isayev (2017): Isayev, E. *Migration, Mobility and Place in Ancient Italy.* Cambridge: Cambridge University Press.

Jacoby (1949): Jacoby, Felix. *Atthis: The Local Chronicles of Ancient Athens.* Oxford: Clarendon.

Jaffee (2001): Jaffee, Martin S. *Torah of the Mouth: Writing and Oral Tradition in Palestinian Judaism, 200 BCE – 400 CE.* Oxford: Oxford University Press.

Jaffee (1991): Janowitz, N. 'The Rhetoric of Translation: Three Early Perspectives on Translating Torah'. *HTR* 84, 129–40.

Janse (2002): Janse, M. 'Aspects of Bilingualism in the History of Greek Language'. In J.N. Adams, M. Janse, and S. Swain (eds.), *Bilingualism in Ancient Society: Language Contact and the Written Text.* Oxford: Oxford University Press, 332–90.

Johnson (2010): Johnson, W. A. *Readers and Reading Culture in the High Roman Empire: A Study of Elite Communities.* Oxford: Oxford University Press.

Johnson/Parker (2009): Johnson, W.A. and H.N. Parker, editors. *Ancient Literacies. The Culture of Reading in Greece and Rome.* Oxford: Oxford University Press.

Jones (1986): Jones, Christopher P. 'Suetonius in the Probus of Giorgio Valla'. *HSPh* 90, 245–51.

Jones (2009): Jones, Christopher P. 'Books and Libraries in a newly-discovered treatise of Galen'. *Journal of Roman Archaeology* 22, 390–97.

Joosten (2012): Joosten, Jan. *Collected Studies on the Septuagint: From Language to Interpretation and Beyond.* FAT 83. Tübingen: Mohr Siebeck.

Kaiser (2010): Kaiser, Stefan. *Die Fragmente des Aristoxenos aus Tarent.* Hildesheim: Olms.

Kaldellis/López Ruiz (2009): Kaldellis, Anthony and Carolina López Ruiz. 'Philon (790)'. *Brill's New Jacoby.* Edited by Ian Worthington.

Kamesar (2009): Kamesar, A. 'Biblical Interpretation in Philo'. In A. Kamesar (ed.), *The Cambridge Companion to Philo.* Cambridge: Cambridge University Press, 65–94.

Kaster (1988): Kaster, Robert. *Guardians of Language: The Grammarian and Society in Late Antiquity.* Berkeley: University of California Press.

Kaster (1995): Kaster, Robert. *C. Suetonius Tranquillus: De Grammaticis et Rhetoribus.* Oxford: Clarendon Press.

Kelly (1979): Kelly, L. *The True Interpreter.* New York: St. Martin's.

Kissel (1990): Kissel, Walter. *Persius Flaccus, Aules, Satyren.* Heidelberg: Winter Verlag.

Knauer (1993): Knauer, E. R. 'Roman wall paintings from Boscotrecase: Three studies in the relationship between writing and painting'. *Metropolitan Museum Journal* 28, 13–46.

König (2008): König, Jason. 'Sympotic dialogue in the first to fifth centuries C.E.'. In Simon Goldhill (ed.), *The End of Dialogue in Antiquity.* Cambridge: Cambridge University Press, 85–114.

König/Woolf (2013a): König, Jason and Gregg Woolf. 'Introduction'. In Jason König and Gregg Woolf (eds.), *Encyclopaedism from Antiquity to the Renaissance.* Cambridge: Cambridge University Press, 1–20.

König/Woolf (2013b): König, Jason and Gregg Woolf. 'Encyclopaedism in the Roman Empire'. In Jason König and Gregg Woolf (eds.), *Encyclopaedism from Antiquity to the Renaissance.* Cambridge: Cambridge University Press, 23–63.

van der Kooij (1998): van der Kooij, A. 'Perspectives of the Septuagint: Who are the Translators?'. In F. García Martínez and E. Noort (eds.), *Perspectives in the Study of the Old Testament and Early Judaism.* Leiden: Brill, 214–29.

Körtge (1901): Körtge, Gustav. *In Suetonii de viris illustribus inquisitionum capita tria.* Dissertationes philologicae Halenses, 14. Halle: Academia Fridericiana.

Kraemer (1996): Kraemer, David. *Reading the Rabbis: The Talmud as Literature.* Oxford: Oxford University Press.

Kramer (2001): Kramer, J. *Glossaria bilinguia altera.* Munich: K.G. Saur.

Kroll (1977): Kroll, John H. 'An archive of the Athenian cavalry'. *Hesperia* 46, 83–140.

Kühn (1821–1833): Kühn, K. G., edited. *Claudii Galeni Opera Omnia.* Leipzig: C. Cnobloch.

Lallot (1998): Lallot, J. *La grammaire de Denys le Thrace: Traduite et annotée.* Second edition. Paris: CNRS Editions.

Lambert (2016): Lambert, Stephen. 'The selective inscribing of laws and decrees in late classical Athens'. *Hyperboreus. Studia classica* 22, 217–39.

Lamberti (1993): Lamberti, Francesca. *"Tabulae Irnitanae": Municipalità e "ius Romanorum".* Naples: Jovene.

Lambrinudakis/Wörrle (1983): Lambrinudakis, Wassilios and Michael Wörrle. 'Ein hellenistisches Reformgesetz über das öffentliche Urkundenweisen von Paros'. *Chiron* 13, 283–368.

Lane Fox (1994): Lane Fox, Robin. 'Aeschines and the Athenian democracy'. In Robin Osborne and Simon Hornblower (eds.), *Ritual, Finance, Politics: Athenian democratic accounts presented to David Lewis.* Oxford: Oxford University Press, 135–56.

de Lange (2015): de Lange, N. *Japheth in the Tents of Shem: Greek Bible Translations in Byzantine Judaism.* Texts and Studies in Medieval and Early Modern Judaism 30. Tübingen: Mohr Siebeck.

Lapin (1999): Lapin, Hayim. 'Rabbis and Cities in Later Roman Palestine: The Literary Evidence'. *Journal of Jewish Studies* 50, 187–207.

Lapin (2000): Lapin, Hayim. 'Rabbis and Cities: Some Aspects of the Rabbinic Movement in its Graeco-Roman Environment'. In Peter Schäfer and Catherine Hezser (eds.), *The Talmud Yerushalmi and Graeco-Roman Culture*, vol. 2. Tübingen: Mohr Siebeck, 51–80.

Lapin (2012): Lapin, Hayim. *Rabbis As Romans: The Rabbinic Movement in Palestine, 100–400 CE.* Oxford: Oxford University Press.

Larson (2011): Larson, J. 'Bilingual Inscriptions and Translation in the Ancient Mediterranean World'. In S. McElduff and E. Sciarrino (eds.), *Complicating the History of Western Translation: The Ancient Mediterranean in Perspective.* Manchester: St. Jerome, 50–61.
Latte (1925): Latte, K. 'Glossographica'. *Philologus* 80, 136–75.
Latte (1953): Latte, K. 'Prolegomena'. In *Hesychii Alexandrini Lexicon* vol. 1. Copenhagen: Munksgaard.
Laurence (2007): Laurence, R. *Roman Pompeii: Space and Society. Second* edition. New York: Routledge.
Laurence/Newsome (2011): Laurence, R. and D. J. Newsome, editors. *Rome, Pompeii, Ostia: Movement and Space.* Oxford: Oxford University Press.
Law/Sluiter (1995): Law, V. and I. Sluiter (eds.), *Dionysius Thrax and the Technē Grammatikē.* Münster: Nodus Publikationen.
Lefèvre (2014): Lefèvre, Eckard. 'P. Terentius Afer'. In Werner Suerbaum (ed.), *Nouvelle histoire de la littérature latine.* Vol. 1: *Des origines à la mort de Sylla.* Turnhout: Brepols, 243–66.
Lefkowitz (2012): Lefkowitz, Mary. *The Lives of the Greek Poets.* London: Bloomsbury.
Lemaire (2006): Lemaire, A. 'Lire, écrire, étudier à Qoumrân et aillerus'. In A. Lemaire and S.C. Mimouni (eds.), *Qoumrân et le judaïsme du tournant de notre ère: actes de la Table ronde, Collège de France, 16 Novenbre 2004.* Leuven: Peeters, 63–79.
Lentz (1867): Lentz, A. *Herodiani technici reliquiae.* Leipzig: Teubner.
Leo (1901): Leo, Friedrich. *Die griechisch-römische Biographie.* Leipzig: Teubner.
Levine (1989): Levine, Lee I. *The Rabbinic Class of Roman Palestine in Late Antiquity.* Jerusalem: The Jewish Theological Seminary.
Lianeri/Zajko (2008): Lianeri, A., and V. Zajko, editors. *Translation and the Classics: Identity As Change in the History of Culture.* Oxford: Oxford University Press.
Liddell *et al.* (1996): Liddell, H. G., R. Scott, H. S. Jones, R. McKenzie, P. G. W. Glare, and A. A. Thompson, *A Greek-English Lexicon.* Ninth edition with revised supplement. Oxford: Clarendon Press.
Lieberman (1962): Lieberman, Saul. *Hellenism in Jewish Palestine.* Second edition. New York, NY: Jewish Theological Seminary of America.
Lieberman (1965): Lieberman, Saul. *Greek in Jewish Palestine.* Second edition. New York, NY: Jewish Theological Seminary of America.
Lightstone (2002): Lightstone, J. N. *Mishnah and the Social Formation of the Early Rabbinic Guild. A Socio-Rhetorical Approach.* Waterloo, Ontario: Wilfrid Laurier University Press.
Lindsay (1994): Lindsay, Hugh. 'Suetonius as 'ab epistulis' to Hadrian and the Early History of the Imperial Correspondence'. *Historia* 43, 454–68.
Lindsay (1995): Lindsay, Hugh. 'Suetonius on the Character of Horace'. *AUMLA* 83, 69–82.
Lobeck (1837): Lobeck, C. A. *Paralipomena Grammaticae Graecae, I-II.* Lepizig: Weidmann.
Lopez-Ruiz (2008): Lopez-Ruiz, Carolina. 'Hieronymos the Egyptian (787)'. In Ian Worthington (ed.), *Brill's New Jacoby.*
Loyd (1969): Loyd, James. *Books in Suetonius' De Vita Caesarum.* Diss. North Carolina.
Lyne (1971): Lyne, Oliver. 'The Dating of the Ciris'. *CQ* 21, 233–53.
Lyne (1978): Lyne, Oliver. *Ciris: A Poem Attributed to Vergil.* Cambridge: Cambridge University Press.
Lyne (1995): Lyne, Oliver. *Horace: Behind the Public Poetry.* New Haven: Yale University Press.
Machiela (2010): Machiela, D. A. 'Once More, with Feeling: Rewritten Scripture in Ancient Judaism —A Review of Recent Developments'. *JJS* 61, 308–20.
Makrinos (2011): Makrinos, A. 'Paraphrases'. In M. Finkelberg (ed.), *The Homer Encyclopedia.* Oxford: Blackwell, 2.626–27.
Mallon (1952): Mallon, Jean. *Paléographie romaine.* Madrid: Instituto Antonio de Nebrija de Filologia.

Mallon (1982): Mallon, Jean. *De l'écriture. Recueil d'études publiées de 1937 à 1981.* Paris: Centre nationale de la recherche scientifique.
Mantovani (2015): Mantovani, Dario. 'Tituli e capita nelle Institutiones di Gaio e nell'Epitome Gai. Contributo allo studio del paratesto negli scritti dei giuristi romani'. *Seminario Complutenses de derecho romano* 28, 587–622.
Mantovani (2017): Mantovani, Dario. 'ERC-Project Redhis: A new appreciation of Juristic texts and Patterns of thought in Late Antiquity. Il programma e i primi risultati'. In Sebastian Lohsse, Salvatore Marino, and Pierangelo Buongiorno (eds.), *Texte wiederherstellen, Kontexte rekonstruieren.* Stuttgart: Franz Steiner Verlag, 167–88.
Mantovani (2018): Mantovani, Dario. *Les juristes écrivains de la Rome antique. Les oeuvres des juristes comme littèrature.* Paris: Les Belles Lettres.
Marec/Pflaum (1952): Marec, Ervan, and Hans-Georg Pflaum. 'Nouvelle inscription sur la carrière de Suétone, l'historien'. CRAI 96, 76–85.
Marrou (1956): Marrou, H. I. *A History of Education in Antiquity.* Translated by G.R. Lamb. London: Sheed and Ward.
Martina (1984): Martina, Mario. 'Le vite antiche di Lucano e di Persio'. *CCC* 5, 155–89.
Martínez Manzano (1994): Martínez Manzano, T. *Konstantinos Laskaris: Humanist, Philologe, Lehrer, Kopist.* Hamburg: Arbeitsbereich II, Byzantinistik und Neugriechische Philologie des Instituts für Griechische und Lateinische Philologie der Universität Hamburg.
Mason (2003): Mason, Steve. *Flavius Josephus, Life of Josephus: Translation and Commentary.* Leiden: Brill.
Matthaios (2010): Matthaios, S. 'Lexikographen über die Schulter geschaut: Tradition und Traditionsbruch in der griechischen Lexikographie'. In Marietta Horster and Christiane Reitz (eds.), *Condensed Texts—Condensing Texts.* Stuttgart: Franz Steiner, 165–207.
Matthaios *et al.* (2011): Matthaios, S., F. Montanari, and A. Rengakos, editors. *Ancient Scholarship and Grammar: Archetypes, Concepts and Contexts.* TCSup 8. Berlin: de Gruyter.
Matusova (2015): Matusova, E. *The Meaning of the Letter of Aristeas: In Light of Biblical Interpretation and Grammatical Tradition, and with Reference to its Historical Context.* FRLANT 260. Göttingen: V&R.
Matusova (2018): Matusova, E. 'The Origins of Translation Theory: The LXX among Jewish Greek Writers'. In M. Meiser, M. Geiger, S. Kreuzer, and M. Sigismund (eds.), *Die Septuaginta – Text, Wirkung, Rezeption: 6. Internationale Fachtagung veranstaltet von Septuaginta Deutsch (LXX.D), Wuppertal 20–24 Juli 2016.* Tübingen: Mohr Siebeck, 557–72.
Mazhuga (2003): Mazhuga, Vladimir. 'A quelle époque vivait le grammairien Phocas?' *RPh* 77, 67–77.
Mazzarino (1955): Mazzarino, Antonio. *Grammaticae Romanae Fragmenta aetatis Caesaris.* Turin: Loescher.
Mazzarino (1973–1974): Mazzarino, Antonio. 'Appunti sul metodo II: Intorno all'età e all'opera di Foca'. *Helikon* 13–14, 505–27.
McElduff (2009): McElduff, S. 'Living at the Level of the Word: Cicero's Rejection of the Interpreter as Translator'. *Translation Studies* 2, 133–46.
McElduff (2013): McElduff, S. *Roman Theories of Translation: Surpassing the Source.* RMCS 14. London: Routledge.
McGill (2010): McGill, Scott. 'Another man's miracles: recasting Aelius Donatus in Phocas's Life of Virgil'. In Scott McGill, Cristiana Sogno, and Edward Watts (eds.), *From the Tetrarchs to the Theodosians: Later Roman History and Culture, 284–450 CE.* Cambridge: Cambridge University Press, 153–70.
McNamee (2007): McNamee, Kathleen. *Annotations in Greek and Latin Texts from Egypt.* Chippenham: American Studies in Papyrology.

Mélèze-Modrzejewski (1995): Mélèze-Modrzejewski, J. 'Law and Justice in Ptolemaic Egypt'. In *Legal Documents of the Hellenistic World: Papers from a Seminar Arranged by the Institute of Classical Studies, the Institute of Jewish Studies and the Warburg Institute, University of London, February to May 1986*. London: Warburg Institute, 1–19.

Mendels/Edrie (2011): Mendels, Doron and Ayre Edrei. *Zweierlei Diaspora. Zur Spaltung der antiken jüdischen Welt*. Göttingen: Vandenhoeck & Ruprecht.

ní Mheallaigh (2008): ní Mheallaigh, Karen. 'Pseudo-Documentarianism and the Limits of Ancient Fiction'. *AJP* 129, 403–31.

Miles (2000): Miles, Richard. 'Communicating culture, identity, and power'. In Janet Huskinson (ed.), *Experiencing Rome: Culture, Identity and Power in the Roman Empire*. London: Routledge, 29–62.

Miller (1895): Miller, Julius. 'Apollonios von Tyana [98]'. *RE* 2.1. 146–48.

Moioli (1987): Moioli, Maria Lauretta. 'La famiglia di Sarapion alias Apollonianus stratego dei nomi Arsinoites ed Hermopolites'. *Acme* 40, 123–36.

Montanari (2015a): Montanari, Franco. '*Ekdosis*. A product of the ancient scholarship'. In Franco Montanari, Stephanos Matthaios and Antonios Rengakos (eds.), *Brill's Companion to Classical Scholarship*. Leiden: Brill, 641–72.

Montanari (2015b): Montanari, Franco. 'Hellenistic Scholarship'. In Franco Montanari, Stephanos Matthaios and Antonios Rengakos (eds.) *Brill's Companion to Ancient Greek Scholarship*. 2 vols. Leiden: Brill, 1.60–183.

Montanari *et al.* (2015): Montanari, Franco, Stephanos Matthaios, and Antonios Rengakos, editors. *Brill's Companion to Ancient Greek Scholarship*, 2 vols. Leiden and Boston, MA: Brill.

Montanari/Pagani (2011): Montanari, Franco and Lara Pagani, editors. *From Scholars to Scholia: Chapters in the History of Ancient Greek Scholarship*. Berlin: de Gruyter.

Montanari/Perrone (2009): Montanari, F. and S. Perrone, editors. *Fragments of the Past: Ancient Scholarship and Greek Papyri*. TiC 1. Berlin: de Gruyter.

Moore (2015): Moore, S. *Jewish Ethnic Identity and Relations in Hellenistic Egypt: With Walls of Iron?*. SupJSJ 171. Leiden: Brill.

Morgan (1998): Morgan, Teresa. *Literate Education in the Hellenistic and Roman Worlds*. CCS. Cambridge: Cambridge University Press.

Morgan (2013): Morgan, Teresa. 'Encyclopaedias of Virtue? Collections of Sayings and Stories About Wise Men in Greek'. In Jason König and Gregg Woolf (eds.), *Encyclopaedism from Antiquity to the Renaissance*. Cambridge: Cambridge University Press, 108–28.

Mossman (1999): Mossman, J. 'Is the Pen Mightier than the Sword? The Failure of Rhetoric in Plutarch's *Demosthenes*'. *Histos* 3, 77–101.

Most (2003): Most, G. W. 'Violets in Crucibles: Translating, Traducing, Transmuting'. *TAPA* 133, 380–90.

Mugridge (2016): Mugridge, A. *Copying Early Christian Texts: A Study of Scribal Practice*. WUNT 362. Tübingen: Mohr Siebeck.

Mulliez (1992): Mulliez, Dominique. 'Les actes d'affranchissement delphiques'. *Cahiers du Centre Gustave Glotz* 3, 31–44.

Munari (1944): Munari, Franco. *Studi sulla Ciris*. Florence: Accademia d'Italia.

Nachmanson (1918): Nachmanson, Ernst. *Erotiani vocum Hippocraticarum collectio cum fragmentis*. Gothenburg: Eranos.

Naiden (2008): Naiden, Fred S. 'Menander (783)'. In Ian Worthington (ed.), *Brill's New Jacoby*.

Naoumides (1969): Naoumides, M. 'The Fragments of Greek Lexicography in the Papyri'. In *Classical Studies Presented to Ben Edwin Perry by his students and colleagues at the University of Illinois, 1924–60*. Urbana: University of Illinois Press, 181–202.

Nasti (2016): Nasti, Fara. 'Papiniano perduto? Una nuova testimonianza sull'estensione della lex Falcidia alle donationes mortis causa'. *Index* 44, 113–22.

Naumann (1979): Naumann, Heinrich. 'Lücken und Einfügungen in den Dichter-Viten Suetons'. *WS* 92, 151–65.

Naumann (1981): Naumann, Heinrich. 'Suetonius' Life of Virgil: The Present State of the Question'. *HSPh* 85, 185–87.

Neusner (2002): Neusner, Jacob. 'The Mishnah in Roman and Christian Contexts'. In Jacob Neusner and Alan J. Avery-Peck (eds.), *The Mishnah in Contemporary Perspective.* Leiden: Brill, 121–48.

Neusner (2004): Neusner, Jacob. *How Not to Study Judaism. Examples and Counter-Examples*, vol. 1: *Parables, Rabbinic Narratives, Rabbis' Biographies, Rabbis' Disputes.* Lanham, ML: University Press of America.

Nicholls (2010): Nicholls, Matthew C. 'Parchment Codices in a new text of Galen'. *Greece & Rome* 57, 378–86.

Nicholls (2011a): Nicholls, Matthew C. 'Bibliotheca Latina Graecaque: on the possible division of Roman libraries by language'. *Latomus* 327: SIEN *Neronia* VIII, 11–21.

Nicholls (2011b): Nicholls, Matthew C. 'Galen and libraries in the Peri Alupias'. *Journal of Roman Studies* 101, 123–42.

Nicholls (2014): Nicholls, Matthew C. *'A library at Antium?'* In C. K. Rothschild and T. W. Thompson (eds.), *Galen's De Indolentia: essays on a newly discovered letter.* Studien und Texte zu Antike und Christentum 88 Tübingen: Mohr Siebeck, 65–78.

Nickau (1966): Nickau, Klaus. *Ammonii qui dicitur liber de adfinium vocabulorum differentia.* Leipzig: Teubner.

Nicolai (2000): Nicolai, R. 'Review of Jocelyn Penny Small "Wax Tablets of the Mind, Cognitive Studies of Memory and Literacy in Classical Antiquity"'. *Athenaeum* 88, 343–47

Niehoff (2011): Niehoff, M. R. *Jewish Exegesis and Homeric Scholarship in Alexandria.* Cambridge: Cambridge University Press.

Nisbet (2007): Nisbet, Robin. 'Horace: Life and Chronology'. In Stephen Harrison (ed.), *The Cambridge Companion to Horace.* Cambridge: Cambridge University Press, 7–21.

O'Hara (2010): O'Hara, James. 'The Unfinished Aeneid?' In Joseph Farrell and Michael Putnam (eds.), *A Companion to Vergil's Aeneid and its Tradition.* Chichester: Wiley-Blackwell, 96–106.

Orlinsky (1975): Orlinsky, H. M. 'The Septuagint as Holy Writ and the Philosophy of the Translators'. *HUCA* 46, 89–114.

Östenberg *et al.* (2015): Östenberg, I., S. Malmberg and J. Bjørnebye, editors. *The Moving City Processions: Passages and Promenades in Ancient Rome.* London: Bloomsbury.

Packer (1997): Packer, J. *The Forum of Trajan in Rome. A Study of the Monuments, I- III.* Berkeley: University of California Press.

Palombi (1990): Palombi, D. 'Gli *horrea* della Via Sacra: degli apunti di G. Boni ad un ipotesi su Nerone'. *Dialoghi di archeologia* 8, 53–72.

Papaconstantinou (2010): Papaconstantinou, A. 'Introduction'. In A. Papaconstantinou (ed.), *The Multilingual Experience in Egypt, from the Ptolemies to the Abbasids.* Farnham: Ashgate, 1–16.

Paratore (1946): Paratore, Ettore. *Una nuova ricostruzione del "De Poetis" di Suetonio.* Rome: Casa editrice Gismondi.

Pedech (1964): Pedech, Paul. *La méthode historique de Polybe.* Paris: Les Belles Lettres.

Pelletier (1962): Pelletier, A. *Lettre d'Aristée à Philocrate.* Sources chrétiennes 89. Paris: Éditions du Cerf.

Peirano (2012): Peirano, Irene. *The Rhetoric of the Roman Fake: Latin Pseudepigraphia in Context.* Cambridge: Cambridge University Press.

Pfeiffer (1949): Pfeiffer, Rudolf. *Callimachus*, 1: *Fragmenta.* Oxford: Clarendon Press.

Pfeiffer (1968): Pfeiffer, Rudolf. *History of Classical Scholarship: from the beginnings to the end of the Hellenistic age.* Oxford: Oxford University Press.
Piccolomini (1879): Piccolomini, E. 'Estratti inediti dei codici greci della bibliotheca Mediceo-Laurentiana, I. Scolii alle orazioni di Gregorio Nazanzeno'. *Annali delle Università Toscane* 17, 231–78.
Pietersma (2002): Pietersma, A. 'A New Paradigm for Addressing an Old Question: The Relevance of the Interlinear Model for the Study of the Septuagint'. In Johann Cook (ed.), *Bible and Computer. The Stellenbosch AIBI-6 Conference: Proceedings of the Association Internationale Bible et Informatique "From Alpha to Byte". 17–21 July, 2000.* Leiden: Brill, 337–64.
Platthy (1968): Platthy, J. *Sources on the Earliest Greek Libraries with Testimonia.* Amsterdam, Hakkert.
Powell (1995): Powell, J. G. F. 'Cicero's translations from the Greek'. In *Cicero the Philosopher: Twelve Papers.* Edited by J.G.F. Powell, pages 273–300. Oxford: Clarendon Press.
Power (2010): Power, Tristan. 'Pliny, Letters 5.10 and the Literary Career of Suetonius'. *JRS* 100, 140–62.
Power (2014): Power, Tristan. 'Juvenal, Satires 3.74 and Suetonius'. *CW* 107, 399–403.
Power (2016): Power, Tristan. 'Poetry and Fiction in Suetonius' Illustrious Men'. In Koen De Temmeman and Kristoffel Demoen (eds.), *Writing Biography in Greece and Rome: Narrative Technique and Fictionalization.* Cambridge: Cambridge University Press, 217–39.
Probert (2006): Probert, Philomen. *Ancient Greek Accentuation.* Oxford: Oxford University Press.
Probert (2011): Probert, Philomen. 'Attic irregularities: their reinterpretation in the light of Atticism'. In Stephanos Matthaios, Franco Montanari, and Antonios Rengakos (eds.), *Ancient Scholarship and Grammar.* Berlin: De Gruyter, 269–90.
Purcell (1993): Purcell, N. 'Atrium Libertatis'. *Papers of the British School at Rome* 61, 125–55.
Quack (2005): Quack, J. F. *Einführung in die altägyptische Literaturgeschichte III: Die demotische und gräko-ägyptische Literatur.* Munster: LIT Verlag.
Radt (2002–2005): Radt, S. *Strabons Geographica.* 4 vols. Göttingen: Vandenhoeck & Ruprecht.
Rajak (2009): Rajak, T. *Translation and Survival: The Greek Bible of the Ancient Jewish Diaspora.* Oxford: Oxford University Press.
Reed (2008): Reed, Annette Yoshiko. "Jewish Christianity' as Counter-History? The Apostolic Past in Eusebius' Ecclesiastical History and the Pseudo-Clementine Homilies'. In Gregg Gardner and Kevin L. Osterloh (eds.), *Antiquity in Antiquity. Jewish and Christian Pasts in the Greco-Roman World.* Tübingen: Mohr Siebeck, 173–216.
Reeve (1984a): Reeve, Michael. 'Terence'. In Leighton Reynolds (ed.), *Texts and Transmission.* Oxford: Clarendon Press, 413–20.
Reeve (1984b): Reeve, Michael. 'The Addressee of the "Laus Pisonis"'. *ICS* 9, 42–48.
Reeve (2011): Reeve, Michael. 'The Vita Plinii'. In Roy Gibson and Ruth Morello (eds.), *Pliny the Elder: Themes and Contexts.* Leiden: Brill, 207–22.
Regenbogen (1950): Regenbogen, Otto. 'Pinax'. *RE* 20.2, 1408–82.
Regenbogen (1860): Reifferscheid, August. *C. Suetoni Tranquilli praeter Caesarum libros reliquiae.* Leipzig: Teubner.
Rhodes (2004): Rhodes, Peter J. 'The laws of Athens in the Aristotelian Athenaion Politeia'. In Delfim F. Leão, Livio Rosetti and Maria do Céu Fialho (eds.), *Nomos. Direito e sociedade na Antiguidade Classica.* Madrid: Ediciones Clasicas, 75–87.
Rhodes (2007): Rhodes, Peter J. 'Documents and the Greek historians'. In John Marincola (ed.), *A Companion to Greek and Roman Historiography.* London: Blackwell, 56–66.
Rhodes (2014): Rhodes, Peter J. *Atthis: The ancient histories of Athens.* Heidelberg: VerlagAntike.
Richardson (1992): Richardson Jr., L. *A New Topographical Dictionary of Ancient Rome.* Baltimore: Johns Hopkins University Press.

Robbins (2002): Robbins, Vernon K. 'Appendix: A Comparison of Mishnah Gittin 1:1–2:2 and James 2:1–13 from the Perspective of Greco-Roman Rhetorical Elaboration'. In Jack N. Lightstone (ed.), *Mishnah and the Social Formation of the Early Rabbinic Guild. A Socio-Rhetorical Approach.* Waterloo, Ontario: Wilfrid Laurier University Press, 201–16.

Robert (1961): Robert, Louis. 'Les épigraphies et l'épigraphie grecque et romaine'. In C. Samarran (ed.), *L'histoire et ses méthodes.* Paris: Pléiade, 453–97.

Robertson (1976): Robertson, Noel. 'False documents at Athens: fifth-century history and fourth-century publicists'. *Historical Reflections* 3, 3–24.

Rochette (1996): Rochette, B. 'Sur le bilinguisme dans l'Égypte gréco-romaine'. *ChrEg* 71, 153–68.

Rochette (1997): Rochette, B. *Le latin dans le monde grec: recherches sur la diffusion de la langue et des lettres latines dans les provinces hellénophones de l'Empire romain.* Brussels: Latomus.

Rostagni (1935): Rostagni, Augusto. 'La Vita suetoniana di Tibullo e la constituzione del Corpus Tibullianum'. *RFIC* 13, 20–51.

Rostagni (1944): Rostagni, Augusto. *Suetonio: De poetis e biografi minori.* Turin: Chiantore.

Roth (1858): Roth, Karl. *C. Suetoni Tranquilli quae supersunt omnia.* Leipzig: Teubner.

Rouse/Rouse (2000): Rouse, R.H. and M.A. Rouse. *Manuscripts and their Makers: Commercial book producers in medieval Paris 1200–1500.* Turnhout: Brepols.

Roussou (2018): Roussou, S. *Pseudo-Arcadius' Epitome of Herodian's De Prosodia Catholica.* Oxford: Oxford University Press.

Rubenstein (2003): Rubenstein, Jeffrey L. *The Culture of the Babylonian Talmud.* Baltimore, MD: The Johns Hopkins University Press.

Rudd (1989): Rudd, Niall. *Horace, Epistles Book II and Epistle to the Pisones ('Ars Poetica').* Cambridge: Cambridge University Press.

Rutherford (2016): Rutherford, I. 'Introduction: Interaction and Translation between Greek Literature and Egypt'. In I. Rutherford (ed.), *Greco-Egyptian Interactions: Literature, Translation, and Culture 500 BCE-300 CE.* Oxford: Oxford University Press, 1–39.

Salanitro (1938): Salanitro, Nino. *Tibullo.* Naples: Casa editrice Luigi Loffredo.

Salati (2018): Salati, Ornella. 'Questioni di diritto militare in un inedito papiro londinese: P.Lond. inv. 2585r e P.Fay. 10 + P.Berol. inv. 11533'. *ZPE* 205, 228–46.

Saldarini (1994): Saldarini, Anthony J. *Matthew's Christian-Jewish Community.* Chicago, IL: University of Chicago Press.

Salmenkivi (2002): Salmenkivi, Erja. *Cartonnage Papyri in Context: New Ptolemaic Documents from Abu Sir al-Malaq.* Helsinki: Societas Scientaiarum Fennica.

Salvadore (1999): Salvadore, Marcello. *M. Terenti Varronis Fragmenta omnia quae extant.* Vol 1. Hildesheim: Olms.

Samely (2002): Samely, Alexander. *Rabbinic Interpretation of Scripture in the Mishnah.* Oxford: Oxford University Press.

Scaffai (1994): Scaffai, M. 'Traduttori latini di Omero'. In R. Degl'Innocenti, S. Orlando and M.P. Pieri (eds.), *La traduzione tra antico e moderno: teoria e prassi; atti del convenvo, Firenze 6–7 dicembre 1991.* Florence: Polistampa, 59–72.

Scafuro (2013): Scafuro, Adele. 'Keeping Record, Making Public. The Epigraphy of Government'. In Hans Beck (ed.), *A Companion to Ancient Greek Government.* London: Wiley, 400–16.

Schäfer (1991): Schäfer, Peter. 'Der vorrabbinische Pharisäismus'. In Martin Hengel and Ulrich Heckel (eds.), *Paulus und das antike Judentum.* Tübingen: Mohr Siebeck, 125–75.

Schäfer (1927–1935): Schanz, Martin and Carl Hosius. *Geschichte der römischen Literatur bis zum Gesetzgebungswerk des Kaisers Justinian*, 2 Vols. Munich: Beck.

Schartau (1994): Schartau, B. *Codices Graeci Haunienses. ein deskriptiver Katalog des griechischen Handschriftenbestandes der Königlichen Bibliothek Kopenhagen.* Copenhagen: Museum Tusculanum Press.

Schironi (2009): Schironi, F. *From Alexandria to Babylon: Near Eastern Languages and Hellenistic Erudition in the Oxyrhynchus Glossary (P.Oxy. 1802 + 4812)*. Sozomena 4. Berlin: de Gruyter.
Schmidt (1922): Schmidt, Friedrich. *Die Pinakes des Kallimachos*. Berlin: Ebering.
Schmidt (1860): Schmidt, M. Ἐπιτομὴ τῆς καθολικῆς προσῳδίας Ἡρωδιανοῦ. Jena: F. Mauk.
Schmidt (1991): Schmidt, Peter. 'Suetons Pratum seit Wessner (1917)'. *ANRW* 2.33.5, 3794–825.
Schmidt (1994): Schmidt, Peter. 'Horaz'. *RAC* 16, 491–524.
Schmidt (2000): Schmidt, Peter. 'C. Suetonius Tranquillus'. In Klaus Sallmann (ed.), *Nouvelle histoire de la littérature latine*. Vol. 4: *L'âge de transition 117–284*. Paris: Brepols, 14–58.
Schofer (2005): Schofer, Jonathan Wyn. *The Making of a Sage: A Study in Rabbinic Ethics*. Madison, WI: University of Wisconsin Press.
Schofield (2008): Schofield, Malcolm. 'Ciceronian dialogue'. In Simon Goldhill (ed.), *The End of Dialogue in Antiquity*. Cambridge: Cambridge University Press, 63–84.
Schon (2004): Schon, Stefan. *Satyros aus Kallatis: Sammlung der Fragmente mit Kommentar*. Basel: Schwabe.
Schwartz (1995): Schwartz, Seth. 'Language, Power and Identity in Ancient Palestine'. *Past and Present* 148, 3–47.
Schwartz (2008): Schwartz, Seth. 'No dialogue at the symposium? Conviviality in Ben Sira and the Palestinian Talmud'. In Simon Goldhill (ed.), *The End of Dialogue in Antiquity*. Cambridge: Cambridge University Press, 193–216.
Sedley (2012): Sedley, D. 'Cicero and the *Timaeus*'. In M. Schofield (ed.), *Aristotle, Plato, and Pythagoreanism in the First Century BC: New Directions for Philosophy*. Cambridge: Cambridge University Press, 187–205.
Shackleton Bailey (1993): Shackleton Bailey, D. R. translator. *Martial. Epigrams, Volume II: Books 6–10*. Loeb Classical Library 95, and *Epigrams, Volume I: Spectacles, Books 1–5*. Loeb Classical Library 94 Cambridge, MA: Harvard University Press.
Shear (1935): Shear, T. L. 'The campaign of 1933'. *Hesperia* 4, 311–39.
Shear (1995): Shear Jr., T. Leslie. 'Bouleuterion, Metroon and the Archives at Athens'. In Mogens H. Hansen and Kurt Raaflaub (eds.), *Studies in the ancient Greek Polis*. Papers from the Copenhagen Polis Centre 6. Stuttgart: Historia Einzelschriften, 157–90.
Shemeh (2009): Shemeh, Aharon. *Halakhah in the Making: The Development of Jewish Law from Qumran to the Rabbis*. Berkeley: University of California Press.
Shutt (1985): Shutt, R. J. H. 'Letter of Aristeas'. In James Charlesworth (ed.), *Old Testament Pseudepigrapha*. 2 vols. New Haven: Yale University Press, 2.7–34.
Sickinger (1999): Sickinger, James. *Public records and archives in classical Athens*. Chapel Hill: University of North Carolina Press.
Sider (2005): Sider, J. D. *The Library of the Villa dei Papyri at Herculaneum*. Los Angeles: J. Paul Getty Museum.
Siebenborn (1976): Siebenborn, Elmar. *Die Lehre von der Sprachrichtigkeit und ihren Kriterien*. Amsterdam: Grüner.
Siméoni (1998): Siméoni, D. 'The Pivotal Status of the Translator's *Habitus*'. *Target* 10, 1–39.
Singer (1997): Singer, P. N. *Galen, Selected Works*. Oxford: Oxford World's Classics.
Sluiter (2000a): Sluiter, Ineke. 'The Rhetoric of Scepticism: Sextus against the Language Specialists'. In Juha Sihvola (ed.), *Ancient Scepticism and the Sceptical Tradition* (*Acta Philosophica Fennica* 66). Helsinki: Philosophical Society of Finland, 93–123.
Sluiter (2000b): Sluiter, Ineke. 'The Dialectics of Genre: Some Aspects of Secondary Literature and Genre in Antiquity'. In Mary Depew and Dirk Obbink (eds.), *Matrices of Genre: Authors, Canons, and Society*. Cambridge, Mass.: Harvard University Press, 183–203.
Sluiter (2011): Sluiter, Ineke. 'A champion of analogy: Herodian's *On lexical singularity*'. In Stephanos Matthaios, Franco Montanari, and Antonios Rengakos (eds.), *Ancient Scholarship and Grammar*. Berlin: De Gruyter, 293–310.

Smith (1978): Smith, O. L. 'Urbano da Belluno and Copenhagen GKS 1965, 4°'. *Scriptorium* 32, 57–59.
Smyth (1956): Smyth, Herbert W. *Greek Grammar.* Revised by G. M. Messing. Cambridge, Mass.: Harvard University Press.
Snyder (2000): Snyder, H. Gregory. *Teachers and Texts in the Ancient World: Philosophers, Jews and Christians.* London: Routledge.
Sommerstein (1991): Sommerstein, Alan H. *The comedies of Aristophanes III: Clouds.* Warminster: Aris & Phillips.
Spanoudakis (2002): Spanoudakis, K. *Philitas of Cos.* Leiden: Brill.
van der Spek (2008): van der Spek, R. J. 'Berossus as a Babylonian chronicler and Greek historian'. In R. J. van der Spek *et al.* (eds.), *Studies in Ancient Near Eastern World View and Society presented to Marten Stol on the occasion of his 65th birthday, 10 November 2005, and his retirement from the Vrije Universiteit Amsterdam.* Bethesda MD: CDL Press, 277–318.
Speyer (1971): Speyer, Wolfgang. *Die litterarische Fälschung im heidnischen und christlichen Altertum: Eine Versuch ihrer Deutung.* Munich: Beck.
Speyer (2002): Spooner, J., editor. *Nine Homeric papyri from Oxyrhynchos.* Florence: Istituto Papirologico "G. Vitelli".
Stasse (2009): Stasse, Baudouin. 'Le traité d'Apamée chez Polybe. Questions de forme'. *Cahiers du Centre Gustave Glotz* 20, 249–63.
Steinby (1999): Steinby, E. M., editor. *Lexicon Topographicum Urbis Romae.* Rome: Edizioni Quasar.
Stem (2005): Stem, Rex. 'Nepos' Atticus as a Biography of Friendship'. *SLLRH* 12, 115–29.
Stenger (2017): Stenger, Jan. 'What Does it Mean to Call the Monasteries of Gaza a 'School'? A Reassessment of Dorotheus' Intellectual Identity'. *Vigiliae Christianae* 71, 59–84.
Stok (2010): Stok, Fabio. 'The Life of Vergil before Donatus'. In Joseph Farrell and Michael Putnam (ed.), *A Companion to Vergil's Aeneid and its Tradition.* Chichester: Wiley-Blackwell, 107–20.
Stronk (2011): Stronk, J. 'Herodotus and Ctesias: Translators of the Oriental Past'. In S. McElduff and E. Sciarrino (eds.) *Complication the History of Western Translation: The Ancient Mediterranean in Perspective.* Manchester: St. Jerome, 101–16.
Stroud (1978): Stroud, Ronald S. 'State documents in archaic Athens'. In *Athens comes of age. From Solon to Salamis.* Princeton: Archaeological Institute of America, 20–42.
Stroud (1979): Stroud, Ronald S. *The Axones and Kyrbeis of Drakon and Solon.* Berkeley: University of California Press.
Strzelecki (1936): Strzelecki, Władysław. 'De Phocae vita Vergiliana'. In *Munera Philologica Ludovico Cwiklinski oblata.* Posnan: Societas Philologiae Polonorum, 235–52.
Swain (1996): Swain, Simon. *Hellenism and Empire: Language, Classicism, and Power in the Greek World, AD 50–250.* Oxford: Clarendon Press.
Swain (2006): Swain, Simon. 'Beyond the Limits of Greek Biography: Galen from Alexandria to the Arabs'. In Brian McGing and Judith Mossman (eds.), *The Limits of Ancient Biography.* Swansea: The Classical Press of Wales, 395–434.
Swartz (1998): Swartz, Michael D. 'Scholasticism as a Comparative Category and the Study of Judaism'. In José Ignacio Cabezon (ed.), *Scholasticism: Cross-Cultural and Comparative Perspectives.* Albany, NY: SUNY Press, 91–114.
Swete (2000): Swete, H. B. *An Introduction to the Old Testament in Greek.* Cambridge: Cambridge University Press.
Syme (1980): Syme, Ronald. 'Guard Prefects of Trajan and Hadrian'. *JRS* 70, 64–80.
Syme (1986): Syme, Ronald. *The Augustan Aristocracy.* Oxford: Clarendon Press.
Tacoma (2016): Tacoma, L. E. *Moving Romans: Migration to Rome in the Principate.* Oxford: Oxford University Press.

Tarrant (1984): Tarrant, Richard. 'Horace'. In Leighton Reynolds (ed.), *Texts and Transmission*. Oxford: Clarendon Press, 182–86.
Tatum (2011): Tatum, Jeffrey. 'The Late Republic: Autobiographies and Memoirs in the Age of the Civil Wars'. In Gabriele Marasco (ed.), *Political Autobiographies and Memoirs in Antiquity*. Leiden: Brill, 161–88.
Tcherikover/Fuks (1957): Tcherikover, V. A., and A. Fuks, editors. *Corpus Papyrorum Judaicarum*. 3 vols. Cambridge, MA: Harvard University Press.
Terzaghi (1931): Terzaghi, Nicola. *Prolegomeni a Terenzio*. Turin: Edizioni de L'Erma.
Thomas (1979): Thomas, Richard. 'Review of Oliver Lyne, *Ciris: A Poem Attribute to Vergil* (Cambridge: Cambridge University Press, 1978)'. *Phoenix* 33, 181–82.
Thompson (2008): Thompson, D. 'The Ptolemaic Library Project'. *Acta Antiqua* 48, 67–72.
Thompson (1966): Thompson, H. A. 'Activity in the Athenian Agora: 1960–1965'. *Hesperia* 35, 37–54.
Thompson/Wycherley (1957): Thompson, H. A., and R.E. Wycherley, *The Athenian Agora III*. Princeton: The American School of Classical Studies at Athens.
Too (2001): Too. Y. L., editor. *Education in Greek and Roman Antiquity*. Leiden: Brill.
Tosi (2015): Tosi, R. 'Typology of Lexicographical Works'. In F. Montanari, S. Matthaios, and A. Rengakos (eds.), *Brill's Companion to Ancient Greek Scholarship*. 2 vols. Leiden: Brill, 1.622–36.
Townend (1961): Townend, Gavin. 'The Hippo Inscription and the Career of Suetonius'. *Historia* 10, 99–109.
Tränkle (2000): Tränkle, Hermann. 'Q. Septimius Florens Tertullianus'. In Klaus Sallmann (ed.), *Nouvelle histoire de la littérature latine*. Vol. 4: *L'âge de transition 117–284*. Paris: Brepols, 494–571.
Trapp (1994:) Trapp, E., W. Hörandner, et al. (eds.) *Lexikon zur Byzantinischen Gräzität*. Vienna.
Tropper (2003): Tropper, Amram. 'Tractate Avot and Early Christian Succession Lists'. In Adam H. Becker and Annette Yoshiko Reed (eds.), *The Ways That Never Parted: Jews and Christians in Late Antiquity and the Early Middle Ages*. Tübingen: Mohr Siebeck, 159–88.
Tucci (2005): Tucci, P. L. 'Where High Moneta leads her steps sublime. The "Tabularium" and the Temple of Juno Moneta'. *Journal of Roman Archaeology* 18, 6–33.
Tucci (2008): Tucci, P. L. 'Galen's storeroom, Rome's libraries, and the fire of A.D. 192'. *Journal of Roman Archaeology* 21, 133–49.
Tucci (2013): Tucci, P. L. 'Flavian libraries in Rome'. In J. König, K. Oikonomopoulou, and G. Woolf (eds.), *Ancient Libraries*. Cambridge: Cambridge University Press, 277–91.
Tuplin (2000): Tuplin, Christopher. 'Nepos and the Origins of Political Biography'. *SLLRH* 10, 124–61.
Turner (1952): Turner, E.G. 'Roman Oxyrhynchus'. *Journal of Egyptian Archaeology* 38, 78–93.
Ucciardello (2012): Ucciardello, G. *Hypomnemata papiracei e lessicografia, tra Alessandria e Bisanzio*. Messina: Università degli studi.
Van Groningen (1963): Van Groningen, B.A. 'ΕΚΔΟΣΙΣ'. *Mnemosyne* 16, 1–17.
Vandorpe (2009): Vandorpe, Katelijn. 'Archives and dossiers'. In Roger S. Bagnall (ed.), *The Oxford Handbook of Papyrology*. Oxford: Oxford University Press, 216–55.
Veenhof (1986): Veenhof, Klaas R. 'Cuneiform archives. An introduction'. In Klaas R. Veenhof (ed.), *Cuneiform Archives and Libraries. Papers read at the 30th Rencontre Assyriologique Internationale. Leiden, 4–8 July 1983*. Leiden: Nederlands historisch-archaeologisch Instituut te Istanbul, 1–36.
Veltri (2006): Veltri, G. *Libraries, Translations, and "Canonic" Texts: The Septuagint, Aquila, and Ben Sira in the Jewish and Christian Traditions*. JSJSup 109. Leiden: Brill.
Vendryes (1945): Vendryes, J. *Traité d' accentuation grecque*. Paris: C. Klincksieck.

Volk (2005): Volk, Katharina. 'Aetna oder Wie man ein Lehrgedicht schreibt'. In Niklas Holzberg (ed.), *Die Appendix Vergiliana: Pseudepigraphen im literarischen Context.* Tübingen: Narr, 68–89.

Walbank (1990): Walbank, Frank W. *Polybius*. Berkeley: University of California Press.

Wallace-Hadrill (1983): Wallace-Hadrill, Andrew. *Suetonius: The Scholar and his Caesars*. London: Duckworth.

Walz (1833): Walz, Christian (ed.) *Rhetores Graeci* vol. V, Stuttgart and Tubingen: J. G. Cotta.

Wardle (2002): Wardle, David. 'Suetonius as *ab epistulis:* An African Connection'. *Historia* 51, 462–80.

Watson (1998): Watson, Alan. *The Digest of Justinian. Volume 2.* Philadelphia: University of Pennsylvania Press.

Watts (2006): Watts, Edward J. *City and School in Late Antique Athens and Alexandria.* Berkeley, CA: University of California Press.

Watts (2012): Watts, Edward J. 'Education: Speaking, Thinking, and Socializing'. In *The Oxford Handbook of Late Antiquity.* Edited by Scott Fitzgerald Johnson, pages 467–86. Oxford: Oxford University Press.

Wehrli (1974): Wehrli, Fritz. *Die Schule des Aristoteles. Supplementband I: Hermippos der Kallimacheer.* Basel: Schwabe.

Welkenhuysen (1968): Welkenhuysen, Andreas. 'Document juridique ou exercise d'écriture? Le P.Mich. VII 431 reconsidéré'. In Willy Peremans (ed.), *Antidorum W. Peremans sexagenario ab alumnis oblatum.* Louvain: Louvain University Press, 309–16.

West (1969): West, S. 'The Greek Version of the Letter of Tefnut'. *JEA* 55, 161–83.

West (1990): West, Martin L. 'Colloquialism and Naive Style in Aeschylus'. In E. M. Craik (ed.), *Owls to Athens: Essays on Classical Subjects Presented to Sir Kenneth Dover.* Oxford: Oxford University Press, 3–12. Reprinted in M. L. West (ed.), *Hellenica* II. Oxford: Oxford University Press, 2013, 203–14.

White (1993): White, Peter. *Promised Verse: Poets in the Society of Augustan Rome.* Cambridge, Mass.: Harvard University Press.

White (2009): White, Peter. 'Bookshops in the Literary Culture of Rome'. In W.A. Johnson and H.N. Parker (eds.), *Ancient Literacies. The Culture of Reading in Greece and Rome.* Oxford: Oxford University Press, 268–87.

Whitmarsh (2005): Whitmarsh, Tim. *The Second Sophistic.* Oxford: Oxford University Press.

Wibier (2014): Wibier, Matthijs. 'The topography of the law book: common structures and modes of reading'. In Laura Jansen (ed.), *The Roman Paratext: Frames, Texts, Readers.* Cambridge: Cambridge University Press, 56–72.

von Wilamowitz-Moellendorff (1921): von Wilamowitz-Moellendorff, U. *Geschichte der Philologie.* Leipzig: Teubner.

Wilson (2007): Wilson, N. G. 'Scholiasts and Commentators'. *Greek, Roman, and Byzantine Studies* 47, 39–70.

Winsbury (2009): Winsbury, R. *The Roman Book: Books, Publishing and Performance in Classical Rome.* London: Bristol Classical Press.

Wise (2015): Wise, M. O. *Language and Literary in Roman Judaea: A Study of the Bar Kokhba Documents.* AYBRL. New Haven: Yale University Press.

Wouters (1979): Wouters, A. *The Grammatical Papyri from Graeco-Roman Egypt: Contributions to the Study of the 'Ars Grammatica' in Antiquity.* Brussels: Paleis der Academiën.

Wouters/Swiggers (2014): Wouters, A. and P. Swiggers, 'Word Classes (mérē toû lógou), Ancient Theories of'. In Georgios Giannakis (ed.), *Encyclopedia of Ancient Greek Language and Linguistics.* Leiden: Brill.

Wright (2003): Wright III, B. G. 'Access to the Source: Cicero, Ben Sira, The Septuagint and Their Audiences'. *JSJ* 34, 1–27.

Wright (2006): Wright III, B. G. '*The Letter of Aristeas* and the Reception History of the Septuagint'. *BIOSCS* 39, 47–67.
Wright (2008): Wright III, B. G. 'Transcribing, Translating, and Interpreting in the *Letter of Aristeas:* On the Nature of the Septuagint'. In A. Voitila and J. Jokiranta (eds.), *Scripture in Translation: Essays on Septuagint, Hebrew Bible and Dead Sea Scrolls in Honour of Rija Sollamo.* JSJSup 126. Leiden: Brill, 147–61.
Wright (2015): Wright III, B. G. *The Letter of Aristeas: 'Aristeas to Philocreates' or 'On the Translation of the Law of the Jews'.* CEJL. Berlin: de Gruyter.
Wycherley (1957): Wycherley, R.E. *The Athenian Agora, Vol. 3, Literary and Epigraphical Testimonia.* Princeton: the American School of Classical Studies.
Xenakis (1969): Xenakis, Jason. *Epictetus Philosopher – Therapist.* The Hague: Martinus Nijhoff.
Xenis (2015a): Xenis, G. *Iohannes Alexandrinus Praecepta Tonica.* Berlin: De Gruyter.
Xenis (2015b): Xenis, G. 'Michael Syncellus: A neglected source for Aelius Herodian's Περὶ καθολικῆς προσῳδίας'. *Classical Quarterly* 65, 868–80.
Yunis (2003): Yunis, Harvey, ed. *Written texts and the rise of literate culture in ancient Greece.* Cambridge: Cambridge University Press.
Zahn (2012): Zahn, M. M. 'Genre and Rewritten Scripture: A Reassessment'. *JBL* 131, 271–88.
Zanker (1995): Zanker, Paul. *The Mask of Socrates: The Image of the Intellectual in Antiquity.* Berkeley: University of California Press.
Ziegler (2016): Ziegler, Joseph. *Sapientia Iesu Filii Sirach.* Septuaginta; Vetus Testamentum Graecum Auctoritate Academiae Litterarum Gottingensis editum vol. XII, 2. Third edition. Göttingen: V&R.
Zuntz (1959): Zuntz, G. 'Aristeas Studies II: Aristeas on the Translation of the Torah'. *JSS* 4, 109–26.

Subject Index

Author Index

Works Cited

Greek Authors

Latin Authors

Jewish

Biblical Texts

Qumran Texts

Rabbinic Literature

Papyri

Inscriptions

Ostraca

Codices